COLLECTIONS OF FROZEN TISSUES

VALUE, MANAGEMENT, FIELD AND LABORATORY PROCEDURES, AND DIRECTORY OF EXISTING COLLECTIONS

Compiled and Edited
by
Herbert C. Dessauer
and
Mark S. Hafner
for the
Workshop Panel

A report to the National Science Foundation by a special workshop panel of tissue collection managers, convened 24–26 May 1983 at the Academy of Natural Sciences of Philadelphia. Sponsored by the Association of Systematics Collections.

Published
by
The Association of Systematics Collections

PREFACE

The Workshop on Frozen Tissue Collection Management, sponsored by the National Science Foundation, convened on May 24, 1983 in Philadelphia, Pennsylvania in conjunction with the annual meeting of the Association of Systematics Collections held at the Academy of Natural Sciences of Philadelphia. In a plenary session, panel members presented a series of papers stressing the value of frozen tissue collections for studies in basic, applied,and forensic sciences. An open discussion followed this presentation. Closed panel sessions were held on May 25-26, 1983 to establish guidelines for collection management and develop a plan to promote and coordinate use, growth, and funding of tissue collections across the nation. The workshop adjourned at noon on May 26, 1983.

This report, based on workshop proceedings, is addressed to researchers, research administrators, and funding agency officials in all fields of basic, applied, and forensic science who are concerned with the nature and quality of America's frozen tissue collections as a national resource. The report presents guidelines and recommendations developed by the Workshop Panel. The recommendations, which are presented in detail in Section VII, are summarized as follows:

1. Establish a Council of the Association of Systematics Collections charged with implementation of the National Plan for Frozen Tissue Collections.
2. A limited number of existing frozen tissue collections in the U.S. should be designated as FROZEN TISSUE DEPOSITORIES, charged with accepting, preserving, and distributing tissues for the scientific community at large.
3. Collections designated as FROZEN TISSUE DEPOSITORIES should, in general, be managed in accordance with traditional curatorial procedures for systematic collections.
4. Major financial support for designated FROZEN TISSUE DEPOSITORIES should be shared by the home institution and the biological resource programs of the National Science Foundation and the National Institutes of Health.
5. Regulatory agencies of the U.S. government should promote the international exchange of frozen tissues.
6. Inventories of collection holdings and advances in the fields of tissue cryopreservation and tissue collection curation should be publicized for the benefit of all.

Included in this report are sections detailing the value of, and need for, collections of frozen tissues, simple methods for collecting tissues under field conditions, present knowledge on long term stability of tissue components, federal regulations concerning tissue collection and transport, and results of a worldwide survey of tissue collection holdings. We sincerely hope that all tissue collection managers, users and potential users, and officials of all federal, state, and private funding agencies will find some value in this report.

Herbert C. Dessauer
Mark S. Hafner

CONTENTS

ACKNOWLEDGEMENTS

The editors and panel members received valuable information and assistance from many sources. We are most appreciative of the financial support provided by the Systematic Biology Program of the National Science Foundation and the advice and support provided by Program Officers Harold K. Voris and James L. Edwards. Excellent administrative support was kindly provided by Stephen R. Edwards, Executive Director of the Association of Systematics Collections, and his staff members A. Michael Neuner and Winifred R. Kucera.

The Academy of Natural Sciences of Philadelphia graciously provided facilities for the panel meetings; we are particularly indebted to Cynthia Lister who made our visit to the Academy both pleasant and productive. Sheldon I. Guttman of Miami University of Ohio and James Frazier of the National Academy of Sciences attended the workshop sessions and provided useful input. Valuable information on protein stability was provided by James L. Patton of the University of California at Berkeley, whose work was supported by National Science Foundation grant GB-37317X. M. F. Shaffer kindly provided the photograph of G. H. F. Nuttall.

This report would not have been possible without the concern and efforts of the large community of frozen tissue collection managers worldwide. Their support and enthusiasm were continual sources of encouragement; their informative responses to our many queries give substance to this report.

This material is based upon a workshop supported by the National Science Foundation under Grant(s) No. BSR - 8218119. The Foundation provides awards for research and education in the sciences. The awardee is wholly responsible for the conduct of such research and preparation of the results for the publication. The Foundation, therefore, does not assume responsibility for such findings or their interpretation.

Any opinions, findings, conclusions, or recommendations expressed in this publication are those of the author(s) and do not necessarily reflect the views of the National Science Foundation.

George Henry Falkiner Nuttall
(1862–1937)

SECTION I
COLLECTIONS OF FROZEN TISSUES: THEIR VALUE TO SCIENCE AND SOCIETY

INTRODUCTION

The British pathologist G. H. F. Nuttall was perhaps the first scientist to recognize the value of undenatured tissues for biological research. Nuttall used sera and antisera to obtain the evidence upon which his classic book *Blood Immunity and Blood Relationships* is based (Nuttall, 1904). The same blood samples collected by Nuttall were used one half century later by Keilin and Wang (1947) to demonstrate the long term stability of blood proteins. In 1948, Professor Alan A. Boyden and his colleagues at Rutgers University initiated the Serological Museum, the first formally organized collection of undenatured tissues. Professor Keilin of Cambridge University deposited in the Serological Museum serum samples originally used by Nuttall. Many of the pioneering studies that utilized undenatured tissues were published in the *Bulletin of the Serological Museum*.

During the past two decades, biological research has moved with ever-increasing speed toward the molecular level. Biochemically oriented publications are appearing with increased frequency, even in nonbiochemical journals. For example, only 3% of the papers in the 1963 volume of *Copeia,* a major journal of ichthyology and herpetology, included molecular data; in contrast, 14% of the papers in the 1983 volume were based on molecular evidence. The 1963 volume of the *Journal of Mammalogy* contained no biochemically oriented papers; the 1983 volume contained eight such papers. In the course of these studies, many scientists have acquired large holdings of frozen tissues from a wide variety of living organisms (see Section V). Today, these collections constitute a valuable, although largely untapped, biological resource for use in basic and applied research.

Many scientists do not realize that a high percentage of macromolecules are stable for extended periods of cryopreservation (Chapter 6). Those investigators who insist on using fresh, unfrozen tissues in their work could often substitute materials already available in frozen tissue collections. In most instances, this would reduce the time and cost involved in acquiring tissues and allow the investigator to sample a broader variety of taxa collected over a span of many years. The literature abounds with studies that would not have been possible had frozen tissues not been available.

Frozen tissues have served as the primary resource for basic research in evolutionary biology, systematics, genetics, biochemistry, and immunology focused at the molecular level (Chapter 1). In addition, these tissues and antisera raised to proteins of the tissues have proven to be of economic importance to scientists and veterinarians conducting breeding programs in zoological parks (Chapter 2), to researchers interested in monitoring levels of environmental pollutants (Chapter 3), and to police, wildlife personnel, and customs agents confronted with forensic problems involving endangered species and regulated game (Chapter 4). In the future, cryopreserved cell cultures and germ cells from natural populations may provide source material for reestablishing species extinct in the wild (Chapter 5).

Frozen tissues from exotic, threatened, and endangered species must be preserved before human encroachment on their habitats and excessive poaching eliminates them (Myers, 1979). For most of these species, traditionally-preserved materials (skin, skeletons, and fluid specimens) are already present in the collections of natural history museums worldwide. Unfortunately, the permanently denatured proteins and nucleic acids of formalin-fixed specimens are useless for studies at the molecular level. Special efforts are now needed to secure tissue samples from threatened species and from species that are rare and difficult to obtain(see Russell, 1978, *Conservation of Germ Plasm Resources: An Imperative,* National Academy of Science). Each animal collected should, in principle, serve as a source of information at many levels of biological organization, from behavior to DNA sequence.

Even though collections of frozen tissues have proven to be a valuable resource for many fields of basic and applied research, it appears that the overall value of these collections is recognized only by a limited community of scientists worldwide (see Stuessy and Thomson, 1981, *Trends, Priorities and Needs in Systematics Biology,* Association of Systematics Collections, 1981; Richardson, 1982, *Workshop on Frozen Tissue Collections in Museums,* Australian Bureau of Flora and Fauna; and, Wickham, 1982, *Proceedings of the U.S. Strategy Conference on Biological Diversity,* U.S. Dept. of State). Moreover, the many extant collections of frozen tissues in the U.S. operate more or less independently with very little coordination of policy or interchange of ideas. We believe, therefore, that it is timely that we document, not only for the benefit of the community of tissue collection managers, but for the scientific community at large, the importance of frozen tissues and frozen tissue collections.

CHAPTER 1

VALUE OF FROZEN TISSUE COLLECTIONS FOR STUDIES IN EVOLUTIONARY BIOLOGY

Herbert C. Dessauer, Mark S. Hafner, and Morris Goodman

Comparisons of molecular components of animal and plant tissues, especially proteins and nucleic acids, yield evidence on the affinities of these organisms at levels that closely reflect gene activity (for reviews see: Leone, 1964; Bryson and Vogel, 1965; Hawkes, 1968; Sibley, 1969a; Manwell and Baker, 1970; Selander and Johnson, 1973; Wright, 1974; Goodman et al., 1976; Wilson et al., 1977; Sigman and Brazier, 1980; Goodman, 1982; Jensen and Fairbrothers, 1983). Because of this, comparative biochemical studies using undenatured tissues have had major impact in the areas of population genetics, systematics, and genomic evolution. In the course of these evolutionary studies, field biologists have collected tissues from many populations of wild organisms. Today, these large accumulations of tissues (see Chapter 12), usually preserved in the frozen state, continue to serve as a primary resource for studies in evolutionary biology.

The extensive collections of skins and skeletons in our natural history museums, now augmented by collections of frozen tissues, allow modern evolutionary biologists to investigate relationships of organisms using data from all levels of biological organization. The systematist is now able to bring into sharper focus animal and plant relationships at all levels, from population to kingdom. The ability to examine genetic affinities at the molecular level is providing new insight into old evolutionary questions, including genetic differentiation and speciation.

Studies in Population Genetics

The discovery of surprisingly large amounts of genetic polymorphism in natural populations, revealed using biochemical techniques, has had a major impact on the discipline of population genetics. In individuals of most species, approximately five percent of the genetic loci are heterozygous; however, in some insects, fish, and amphibians, heterozygosity levels exceeding 10 percent have been observed (Nevo, 1978). The current "neutralist- selectionist" debate among population geneticists is a direct outgrowth of their attempt to understand the biological meaning of genetic polymorphism in natural populations (Nei, 1975; Nevo, 1978).

Genetic evidence, obtained largely through isozyme analysis, is now available on representatives of virtually all groups of organisms. The remarkably large amounts of genetic variation observed have given us the means for solving many kinds of problems. These include questions concerned with: (a) the genetic uniqueness of the individual; (b) breeding patterns within a population; (c) the genetic parameters that describe the population; (d) the magnitude and nature of genetic interactions in contact zones between members of different populations; (e) the identification of sibling species; and, (f) the estimation of the magnitude of genetic divergence between groups of closely related organisms.

Studies of the genetic diversification of geographically widespread organisms, and studies of retrospective changes in the genetics of populations (see Chapter 3) are especially dependent upon frozen tissue resources. Often years of collecting are required to accumulate sufficient samples to complete a project involving organisms from widely separated geographic areas. In the interim, samples collected must be preserved in the undenatured, usually frozen, state. For example, four years of collecting were required to obtain the samples of spiny lobsters needed to evaluate the influence of ocean currents on gene flow, larval dispersal and recruitment to benthic populations (Menzies, 1981). Lawson and Dessauer's (1979) study of geographic variation in garter snakes utilized tissues collected over a span of two decades; these same tissues, preserved in ultracold freezers, were available for studies in blood chemistry (Dessauer, 1970) and immunology (Mao and Dessauer, 1971). Thus, samples of tissues frozen for extensive periods of time have been used for a wide variety of investigations. A bibliography of the many and varied biochemical studies of natural vertebrate populations is now available (M. W. Smith et al., 1982).

Studies in Biochemical Systematics

Protein electrophoresis (Harris and Hopkinson, 1976) has proven to be a valuable means of studying animal and plant relationships at lower taxonomic levels, usually below the level of the genus. Studies of protein differentiation, calibrated against the fossil record, have revealed the existence of rapidly and slowly evolving classes of proteins (Sarich, 1977). The majority of samples in frozen tissue collections today were collected by systematists working at lower taxonomic levels, using electrophoretic techniques (see Chapter 11).

Although immunological methods were the first techniques used to obtain molecular assessments of the relationships of organisms (Nuttall, 1904), comparative immunology is still widely used in systematic research. Blood, eye, venom, and seed proteins are the antigens most commonly studied; however, tissues in frozen collections are a potential source for a wide variety of other antigens yet to be studied. Most immunological research has involved vertebrates, and especially mammals; however, taxa of virtually all groups of animals (Leone, 1968) and plants (Fairbrothers, 1969) have been examined.

In addition to providing evidence on relationships among higher taxa, comparative immunology has provided much of the data upon which the "molecular clock" hypothesis is based (Wilson et al., 1977). This hypothesis postulates that the rate of accumulation of structural gene mutations is relatively constant over geological time. The "molecular clock" hypothesis has stimulated much research and controversy concerning rates of evolutionary change in macromolecules (Goodman, 1982).

Several techniques for the comparison of the nucleotide sequences of DNA (Miniatis et al., 1982) are being applied to problems in systematics and evolutionary biology. In the DNA hybridization method, the single-stranded homologous nucleotide sequences of different species are "hybridized" to form duplex molecules, the measurable properties of which are proportional to the phylogenetic proximity of the taxa. The data provide an index to the relative times of divergence among living taxa, and thus to the branching pattern of their phylogeny. When calibrated against geological or fossil datings, the DNA hybridization values are believed by some workers to provide measures of absolute time. Britten and Kohne (1968), Kohne (1970), Shield and Straus (1975), and Sibley and Ahlquist (1981b) have described the DNA hybridization technique and its application to systematics. Both sequencing and restriction endonuclease fingerprinting analysis of DNA have been used in studies at the species and intraspecific levels (Shah and Langley, 1979; Brown, 1980; Munjaal et al., 1981; Aquadro and Greenberg, 1983). As the number of systematists using comparative DNA techniques increases, the need for collections of properly preserved sources of DNA will also increase (Sibley and Ahlquist, 1981a, 1983).

Studies on the Structure and Evolution of the Genome

Comparative biochemical studies yield evidence not only on the organisms studied, but also on their evolving genes and proteins (Bryson and Vogel, 1965; Sigman and Brazier, 1980; Fitch, 1982). Comparisons of sequences of proteins and nucleic acids provide the maximum amount of evidence on the structure of the genome. Scientists carrying out such studies are often dependent upon frozen tissue collections. For example, tissues from crocodilians in the collection of the Louisiana State University Museum of Zoology have been sent to investigators in the United States, Germany and Belgium for use in sequencing myoglobin, hemoglobin and alpha-crystallin. Because of the expense and time required to sequence proteins, large comparative data sets are available for only a relatively small number of specific proteins, usually only those available in large quantities. However, this data base will increase rapidly in the near future as molecular biologists clone and sequence additional structural genes (Sanger, 1981).

Genealogical reconstructions using protein and nucleic acid sequence data are providing valuable information on the tempo and mode of evolutionary change. For example, data on proteins such as the globins suggest that accelerated rates of structural change occur early in the history of a lineage and decelerate later (Goodman, 1982). Comparative studies of sequences have revealed the existence of families of genes and proteins, making possible genetic classifications of macromolecules (Dayhoff et al., 1983). Our knowledge of the structure and activity of one family of nucleic acids, the oncogenes, promises to help elucidate the genetics and etiology of carcinogenesis (D'Eustachio, 1984; Gardner et al., 1984; see Chapter 3). Collections of frozen tissues have been, and will continue to be, a major factor in making possible this expansion of our knowledge of the biosphere.

CHAPTER 2

THE VALUE OF FROZEN TISSUE COLLECTIONS FOR ZOOLOGICAL PARKS

Oliver A. Ryder and Kurt Benirschke*

Research has been conducted in zoological gardens from their inception as scientific institutions (as opposed to menageries). However, it has been realized only recently that research in zoos is necessary for the preservation of certain vertebrate species. Zoo-oriented research, including research designed to establish self-sustaining populations, is currently conducted at a high level of sophistication. Investigators in such disciplines as pathology, genetics, ethology, virology, immunology, endocrinology and reproductive physiology are applying modern biomedical techniques to the problem of endangered animal species. At the same time, animals in zoological parks and aquariums, representing the diversity of vertebrate life, are being used for an ever-increasing variety of scientific studies.

The Frozen Zoo

In order to serve a diversity of in-house research needs and to make available specimens of vanishing life forms for scientific inquiry, several zoos in the U.S. have established collections of frozen tissues (Benirschke, 1982; see Chapter 12). The Zoological Society of San Diego has assumed a leadership role in this area.

Researchers at the San Diego Zoo routinely sample animal tissues during necropsy, routine medical examinations, and when animals are ear-notched for identification purposes. Primary fibroblast cultures are established from sterile skin biopsies, and semen samples are collected opportunistically. Many of these tissue and semen samples are cryopreserved. Mammalian embryos may be collected following *in vivo* fertilization of single or multiple ova. Embryos may be immediately transferred to suitable recipients or frozen for later use. The frozen tissue collection of the San Diego Zoo has proven to be a valuable resource for both in-house and outside investigators.

Uses of Frozen Tissues in Zoological Parks

Frozen tissues collected in zoological parks have a wide variety of uses in both basic and applied science. For example, these tissues have been valuable for studies in population genetics, systematics, and evolutionary biology (e.g., Leboffe, 1979; Woodruff and Ryder, in prep.).

Biochemical Genetic Studies.–Knowledge of intraspecific allelic variation can be applied to questions of parentage uncertainty. Thus, when the identity of the father of a newborn bonobo (*Pan paniscus*) at the San Diego Zoo was uncertain, biochemical genetic studies of blood proteins helped establish the identity of the infant's father (Woodruff and Ryder, in prep.) and confirmed evidence based on polymorphic chromosomal markers (Benirschke and Kumamoto, 1983). Similarly, analysis of a globin chain polymorphism of the Arabian oryx identified a case of mistaken maternal

*Research Department, Zoological Society of San Diego

identity and allowed correction of the studbook for this species (Ryder et al., 1981).

Often, breeding groups in zoos consist of small numbers of animals. Over even a few generations considerable genetic variation may be lost due to genetic drift (Senner, 1980). Documentation of this phenomenon has been achieved for the captive population of the Asiatic wild horse, *Equus przewalskii* (Ryder et al., 1982).

Highly repetitive satellite DNAs are an enigma within eukaryotic genomes. The evolution and cytological dynamics of satellite DNAs within the Equidae are being investigated at the San Diego Zoo (Ryder and Hansen, 1979; Gadi and Ryder, 1983) and access to frozen tissue samples of pedigreed individuals as well as separate geographic races or subspecies provides a valuable experimental resource.

In the future, DNA sequencing will provide more detailed insights into molecular evolution than was hitherto possible (see Chapter 1). In some cases, access to DNA from rare species may be available through recombinant genomic libraries (e.g., Asian wild horse and gorilla). In most cases, and indeed, in order to construct genomic libraries, fresh or freshly frozen tissue is required for DNA extraction.

It is usually impossible to anticipate when tissue samples will become available. In order to plan studies and to take advantage of valuable specimens, freezing technology must be employed. In other words, a frozen collection of tissues and fibroblast strains is essential for productive research efforts.

Availability of cell lines from endangered species whose future availability is uncertain insures continued access to raw materials for genetic study. The only feasible method of collecting samples from such species for frozen collections is on-site at zoological gardens.

Cytogenetic Studies.–The availability of cryopreserved fibroblast cultures has greatly aided studies in comparative mammalian cytogenetics. As new techniques become available or as new questions arise the appropriate cultures are thawed, propagated, and chromosome harvests made. A recent retrospective survey of a heterochromatin polymorphism of the Asian wild horse involved over a dozen animals that had died since the time their fibroblasts were frozen (Ryder et al., 1983). Furthermore, and perhaps most importantly, many species of large mammals face probable extinction in the coming years. For such species as the gorilla, red-ruffed lemur, and lowland anoa (*Anoa depressicornis*) future access to tissue may only be available from cryogenically preserved cells.

In establishing captive self-sustaining breeding programs for rare species, it is advantageous to characterize the founder stock as thoroughly as possible. As an example, the orang-utan (*Pongo pygmaeus*) consists of two distinct geographic forms that exhibit clear differences in karyotype and also considerable morphological variability. Unfortunately, not all breeding pairs of wild-caught orang-utans have been matched with respect to their subspecies status. Individuals of uncertain ancestry may be identified as Bornean orang-utans, *Pongo p. pygmaeus,* or Sumatran orang-utans, *P. p. abeli,* or hybrids, by their karyotype. Scientists at the San Diego Zoo are collecting this information to assist them in managing the breeding program of orang-utans and to avoid admixture of the two subspecies' breeding stock.

Chromosomal incompatibilities are a frequent cause of fetal wastage in humans and the same probably holds for other species as well. Habitual abortion in the colony of endangered Douc langurs (*Pygathrix nemaeus*) in the San Diego Zoo was associated with a chromosomal variant in one breeding male (Benirschke and Bogart, 1978).

Health-Related Applications.–The availability of frozen tissues greatly aids investigations into agents of infectious disease and pathology (see Chapter 3). Retrospective studies of a hepatitis-like virus that may be a major cause of liver pathology in exotic felines has depended on frozen samples. Similarly, these collections have been valuable for studies of trace element levels in ungulates at the San Diego Zoo.

Basic research into comparative reproductive physiology is dependent on frozen samples of adrenal, testis, placenta, and other endocrine tissues. For example, utilizing frozen placental tissue, Doellgast and Benirschke (1979) were able to examine levels of alkaline phosphatase in non-human primates and determine that gorillas and spider monkeys have three orders of magnitude less heat-stable alkaline phosphatase than do human, chimpanzee, and orang-utan placentas.

Gene Pool Preservation.–Collections of frozen spermatozoa (see Chapters 5 and 8) are an immensely valuable resource for the genetic management of small populations of endangered species. To minimize genetic changes of captively-bred animals compared to their non-captive relatives, management schemes must attempt to minimize inbreeding, maximize the number of genetically effective individuals, equalize the contribution of founder individuals, and keep large enough population sizes so that a minimum of genetic diversity is lost over time. Thus, there is an increasing incidence of purposefully designed matings between animals located in different zoos. Often, desirable matings are difficult to arrange because the expense of shipping an animal is often prohibitive, represents too great a risk to the health of the individual, or is restricted by government regulations (see Chapter 9). In such cases, artificial insemination using semen preserved in frozen collections may circumvent such difficulties.

The carrying capacity of zoos, i.e., the amount of living space for animals, is far below the space needed to maintain genetically sound breeding groups of the 350 mammalian and 450 avian species that are currently listed as threatened or endangered. One possible solution to this dilemma would be to use cryogenically preserved gametes and embryos to enhance the carrying capacity of zoos. For example, if semen from a wild-caught male snow leopard were to be collected, aliquoted, and cryogenically preserved, it could be used to father offspring generations after the death of the donor. With judicious planning, preserved semen from several males could be used to equalize founder representation and minimize inbreeding in captive populations of snow leopards. Using this approach, fewer living snow leopards would be needed to maintain an acceptable level of genetic variability. This approach has yet to be used in zoological parks, but certainly will be employed in the future. In the interim, considerable research in cryopreservation, reproductive physiology, and genetics will be needed.

Embryo transfers may prove valuable in the future for the propagation of endangered species. When females of an endangered species are too few to produce the number of young prescribed by a breeding program, the use of embryo transfers to suitable recipient females of a related species may prove desirable. Although the theoretical benefits of allospecies embryo transfer are obvious, many practical problems must first be overcome.

Conclusions

Although frozen tissues collected from animals in zoological parks have proven valuable for many kinds of basic and applied research, their potential value has yet to be realized. At present, only a few zoos are collecting and preserving frozen tissues (Benirschke, 1982) and fewer still have frozen cell and gamete collections. Encouragingly, however, there is much cooperation among personnel from different zoos, and zoo professionals are keenly aware of the need for future research involving cryopreserved tissues. The American Association of Zoological Parks and Aquariums is considering how to organize, coordinate, and fund the necessary research. Regrettably, no national or international funding agency exists to underwrite research on captive propagation technology. There is no "NIH for Conservation" and no "National Endowment for the Works of Nature." Nonetheless, progress is being made in those zoological institutions that are financially able to direct resources toward research activities. In summary, the frozen cell and tissue collections of animal species represented in zoological parks are now, and will continue to be, an invaluable asset for basic research in biology and for conservation of vertebrate species.

CHAPTER 3

VALUE OF FROZEN TISSUE COLLECTIONS FOR ENVIRONMENTAL MONITORING AND RETROSPECTIVE STUDIES

Michael H. Smith and Mark S. Hafner

Hindsight, or past experience, is often the only basis for predicting the environmental impact of new programs or projects. More often than not, future plans are predicated on simple empiricism, i.e., if a past program proved to be detrimental, don't repeat it. However, a retrospective study based on environmental sampling through time represents "planned hindsight," allowing one to address a wide variety of historical questions at a more sophisticated level. Samples collected to explore one problem may prove to be of historical value for investigating many future problems in distantly related areas.

Many past environmental conditions can be inferred through analysis of preserved animal and plant tissues. If the tissues were preserved in the frozen state, they can be used to estimate the presence in past environmemts of specific radioactive isotopes, commercial and agricultural chemicals, industrial pollutants, and pathogens. One is also able to determine genotypic frequencies that characterized a population in the past and assess other historical genomic properties, including the presence, structure, and position of particular genes.

Studies of biochemical variation in natural populations can, to some degree, avoid the hindsight dilemma. Frozen tissues collected from specific populations over a long period of time can be used to reanalyze a problem from a new perspective. Analyses can be repeated to compare old with new samples, and the data base can be extended using new techniques as they become available. In the case of environmental pollution, the problems, being ecological, are usually very complex; thus, our ability to ask the critical questions is limited. However, environmental pollution often involves extreme situations such that changes in allele frequencies in response to selection may be expected (Bishop and Cook, 1981; Smith, M. W. et al., 1983). Frozen tissue samples allow us to compare allele frequencies in populations before and after environmental perturbation to seek a possible link between frequency changes and the environmental disturbance.

A study of allele frequency changes in populations of largemouth bass (Smith, M. H. et al., 1983) illustrates the value of frozen tissue collections for retrospective environmental studies. The study was designed to examine allele frequencies in bass populations in thermally altered, post-thermal, and unaltered environments. The bass in post-thermal environments were living in areas that had previously received heated effluents from a nuclear production reactor. These populations had intermediate allele frequencies relative to bass populations living in thermally unaltered environments and those living in locations currently receiving heated effluents. Yardley and colleagues (1974) hypothesized that the post-thermal bass population might be in the process of evolving back to allele frequencies characteristic of populations in unaltered areas. Tissues had been sampled from fish populations throughout the period of the disturbance, and frequency data developed from an analysis of one malate dehydrogenase locus in these tissues supported the hypothesis. Unfortunately, however, tissue samples from bass populations were not collected from the hot water environment when the reactor was first put into operation. In the absence of these critical samples, our understanding of the effects of thermal perturbation on bass populations remains incomplete.

There are an increasing number of examples where frozen tissue collections have proven to be of value in retrospective studies. For example, blood collected 20 years ago for a genetic study of dairy cattle was recently used to investigate the epidemiology of blue tongue virus in Louisiana. A similar collection of armadillo sera, collected between 1958 and 1962 for a study of wildlife leptospirosis, was later shown to be ELIZA-positive for *Mycobacterium leprae* infections. This discovery predates, by approximately eight years, the first known case of leprosy in armadillos (M. E. Hugh-Jones, pers. comm.). Human blood samples taken 10 years ago for a study concerned with heart disease were used recently in a retrospective investigation of the possible association between cancer risk and blood levels of vitamins A and E (Willett and MacMahon, 1984). Samples of lymphoid tissue collected in the early 1970's from a colony of monkeys at the National Cancer Institute proved to be valuable in a recent study of simian acquired immune deficiency syndrome. This disease, caused by a type D retrovirus, is similar in many respects to human AIDS. Studies of the frozen primate tissues enabled investigators to trace the epidemic of simian AIDS back to at least 1976 (Gardner et al., 1984, in press; J. Casey, pers. comm.).

As the above examples illustrate, many retrospective studies today would not be possible were it not for the fortuitous availability of appropriate tissues in frozen collections. Undoubtedly, current collections of frozen tissues (see Chapter 12) already contain critical samples for retrospective studies yet to be designed. For example, collections of tissues from migratory birds may prove valuable in understanding the epidemiology of viral diseases that affect domestic poultry, and tissues collected in the past from many species will be of obvious value in future retrospective studies of pesticide residues.

Although there will continue to be occasional fortuitous uses for frozen tissues, their value in retrospective studies should increase dramatically with carefully planned sampling programs. It is obvious that samples of animal and plant tissues should be preserved prior to major environmental perturbations. For example, investigators should not miss the opportunity to collect baseline samples of tissues from key species before nuclear reactors, waste disposal sites, industrial plants, pesticide treatments, and other large scale programs known to affect the environment are placed into operation.

With reasonable foresight, future environmental monitoring programs will become more purposeful, economically feasible, and scientifically sound. Frozen tissue collections should play an increasingly important role in these endeavors.

CHAPTER 4

VALUE OF FROZEN TISSUE COLLECTIONS FOR FORENSIC STUDIES

Herbert C. Dessauer and Kenneth W. Goddard

Frozen tissue collections are of considerable value to agents of the U.S. Fish and Wildlife Service and related state agencies charged with the protection of wildlife and the prevention of illicit trafficking in wildlife parts and products. In their daily tasks, agents are often in need of species identifications of animals or parts of animals taken as evidence. Museum specialists, including ornithologists, mammalogists and the like, are usually able to identify whole animals, feathers or bones; however, agents are occasionally faced with the difficult task of identifying the species source of evidence that may consist of only a blood stain or tissue fragment. Such evidence often contains species-specific biochemical markers that may be stable for long periods of time (see Chapter 6), even under the adverse conditions usually associated with criminal evidence. In these cases, a properly trained scientist may be able to identify the source of such material by means of comparative protein analysis (Davies, 1975). With proper comparative material, the modern forensic scientist should be able to identify the species source of animal parts and products, even when all external species-distinguishing characters have been destroyed. There are many case histories where biochemical methods were applied to forensic problems; however, the majority of these are not reported in the scientific literature. Two pioneering studies illustrating the value of this approach are those of Lillevick and Schloemer (1961), who were able to identify mislabeled fish filets in a commercial market, and Baker (1966), who identified tissue fragments from birds that had collided with a jet aircraft.

Although molecular methods are extremely powerful means for species identification, the results are almost invariably challenged by the legal community when offered in evidence. Therefore, the forensic scientist must often be able to defend his findings under intense legal cross-examination. Not only must he be an expert in the methodology used, but he must also convince the court that his findings were based upon the use of properly identified reference standards. These standards, include frozen tissue collections consisting of undenatured tissue samples and species-specific antisera. To meet court requirements, it is important that standards be preserved properly and that the organism from which they were collected be correctly identified (with locality and date of capture recorded). Preferably, the species source of the standard should be documented in the form of a traditional museum voucher specimen (see Section III).

Many wildlife agents faced with forensic problems are unaware of the existence in the U.S. of many frozen tissue collections that could serve as sources for forensic standards (see Section IV). In addition, they are unaware of the wide variety of skills possessed by collection managers in the area of species identification using molecular methods.

The importance of increased communication between wildlife agents and scientists managing frozen tissue collections is exemplified by a recent problem encountered by agents in Brownsville, Texas. A large shipment of meat suspected of being from the endangered Olive Ridley sea turtle was seized. Fortunately, agents were aware of the turtle research conducted by Dr. Robert A. Menzies, manager of the tissue collection at Nova University. When informed of the problem, Menzies organized the efforts of the scientific community to address the issue. Biochemical and immunological analyses of the seized material were carried out at Nova University, the National Marine Fisheries Service Charleston Laboratory, and the Louisiana State University Medical Center using samples in their frozen collections as standards. Specific antisera for the immunological work was obtained from the collection at The King's College in New York. The biochemical and immunological comparisons proved that the seized meat was indeed from Olive Ridley sea turtles.

Police crime labs at the federal, state, and local levels are generally overburdened with human-related evidence (Davies, 1975) and usually cannot devote the effort and resources needed to develop wildlife-related identifications. Furthermore, they usually do not have the tissues and specific antisera necessary to make critical identifications. For example, the manager of the LSU Museum of Zoology frozen tissue collection was able to correctly identify a sample of blood found in a boat belonging to a hunter suspected of poaching alligators. The wildlife agent involved first took the evidence to a Parish forensic lab where it was incorrectly identified because of the lack of undenatured alligator tissue standards and inactive antisera. Because the agent was convinced of the hunter's guilt, he sought a second opinion. Fortunately, the LSU collection included appropriate reference tissue samples and potent antisera to alligator proteins, enabling unequivocal identification of the source of the evidence.

The U.S. Fish and Wildlife service program for the prevention of illicit trafficking in wildlife parts and products will certainly be dependent upon frozen collections of tissues and antisera for standards necessary to identify the source of seized wildlife tissues. In the future, forensic specialists from the Service's Law Enforcement Forensic Branch hope to be conducting extensive searches for documented wildlife parts and products to serve as research and casework standards (e.g., blood, tissues, hair, fur, hides, teeth, claws, and the like). A substantial number of sources for undenatured tissue, blood and antisera reference samples have already been located (Section IV); many others probably exist.

CHAPTER 5

VALUE OF FROZEN TISSUE COLLECTIONS FOR GENE POOL PRESERVATION

George F. Gee

A gene pool, for the purpose of preservation, is a collection of living organisms, parts, or cells capable of regenerating the population from which it came. Historically, animal gene pools have been preserved in the form of domestic and nondomestic animal collections (e.g., farm herds and breeding populations in zoological parks) and by the protection of free-living species in their native habitat (e.g., wildlife preserves and refuges). However, it is impossible to captively propagate or manage large numbers of animal species because of the vast number of breeding programs necessary, the area and number of refuges required, and the costs of these programs.

Importantly, an animal population, whether in captivity or on a refuge, must be large enough to maintain its vitality, which is a function of its genetic diversity (Terborgh and Winter, 1980); once genetic diversity is lost in a species, there is no way to restore it. Conway (1980) estimated that all the zoos in the world have the combined capacity to maintain adequate genetic diversity in only 100 species of animals. Preservation of diversity assures continued richness of the earth's biota, and in the future may even be necessary for man's survival (see Russell, 1978, *Conservation of Germ Plasm Resources: An Imperative*). Myers (1979) predicted that up to two million of the estimated 10 million species of plants, animals, and microorganisms in the world today will be extinct by the year 2000.

Cryopreservation

Fortunately, modern methods of tissue preservation offer a simple and relatively inexpensive means of preserving genetic resources. Cryopreservation techniques have made it feasible to store tissues, blood, and gametes and to thaw them when needed with little loss in viability (see Chapter 6). Cryopreservation of gametes and embryos makes it possible to maintain the integrity of a species during periods of great risk or prolonged restoration efforts. Using cryopreservation techniques, one is now able to collect, preserve, and implant embryos in certain mammalian species, and to collect and preserve sperm for artificial insemination of many other vertebrate species (Gee and Sexton, 1979; Gee and Temple, 1978; Polge, 1978; Seager et al., 1978).

Cryopreservation methods are useful for preserving tissues from a wide range of animal and plant species. The discovery of the cryoprotectant properties of glycerol and dimethylsulfoxide (DMSO), and the development of freezing devices during the past 40 years, has propelled cryobiology from a laboratory tool to a commercial enterprise. Human and cattle semen banks and human blood banks are examples of applied cryogenic developments. In addition, techniques have been developed for successful cryopreservation of semen from other domestic and nondomestic species (currently, 40 to 50 mammalian species, and several reptile and bird species; Sexton and Gee, 1978; Graham et al., 1978; Seager et al., 1978; Purcel, 1979).

Practical Frozen Gene Pools

Much additional research on freezing and storing gametes and embryos is needed if germ plasm cryopreservation is to become a viable management tool. Workshops to evaluate the potential benefits of new methods of gene pool conservation are of great benefit to the endangered species preservation effort and the conservation of our natural resources. Cryogenics will play a major role in future conservation programs because of its potential for widespread use, adaptability, and cost effectiveness.

Living cells must be handled with extreme care if they are to retain their viability during cryopreservation (see Chapters 6 and 8). Of the three candidates for a frozen gene pool (ova, sperm, and embryos), ova preservation poses the most serious technical difficulties during collection, life support, freezing, fertilization, and transfer to host animals; practical methods of ova preservation have yet to be developed. Cryogenic preservation of whole embryos is limited to dairy cattle and a few species of laboratory animals (Seidel, 1981). Embryo preservation has an important advantage over gamete preservation in that stored embryos are whole animals (diploid) and not germ cells (haploid). Thus, embryos of endangered species can be transferred to non-threatened animals for implantation, gestation, and rearing. Embryo transfer frees the endangered animals from the stress of pregnancy and increases the number of offspring produced (Waladsen et al., 1977; Seidel, 1981). For example, in 1977 at the University of Utah, the embryo of a wild Sardinian sheep was transferred to the uterus of a domestic sheep, where it developed into a healthy lamb. A similar transfer of a gaur embryo to a Holstein cow was successful at the Bronx Zoo (J. Stover, pers. comm.).

Frozen Semen

At present, semen cryopreservation is the most practical procedure for the preservation of gene pools. Semen storage at liquid nitrogen temperatures (-196°C) appears to stop all metabolic activity, and the reproductive competence of such semen remains high for extremely long periods of time, possibly for centuries. Cryopreservation of semen from several mammalian species, including dairy cattle, was practical by 1960. Drs. Seager, Wildt, and Platz (1978) of the National Institutes of Health and Texas A&M University developed a frozen semen bank for many nondomestic animals and demonstrated the competence of the thawed semen. Avian semen preservation was not practical until the mid-1970's when domestic fowl were successfully inseminated with thawed semen cryoprotected by DMSO at the Beltsville Agricultural Research Center (Sexton, 1976) and with thawed semen cryoprotected by glycerol at other centers (Lake, 1978). In 1976, the Patuxent Wildlife Research Center and the Beltsville group began successful cooperative research on avian artificial insemination, first with cranes, and later with ducks and geese.

Although cryopreservation of semen has been successful for many species of vertebrates, each species presents unique problems that must be solved (Graham et al., 1978). For example, in certain species semen from some individual animals does not preserve as well as that from others in the population, and breeding programs in gene pool preservation projects should be designed to include many individuals, even if the fertility rates are lower for a few. A greater number of samples must be stored from an animal whose expected fertility rate is 30% than from another whose expected fertility rate is 80%. The number of individual animals from each species to be stored in a semen gene pool is a topic of discussion at many professional conferences, and is subject to many variables. The optimal number is obviously greater than 10, but for certain endangered species 10 or fewer individuals may be all that remain. Obviously, the number of individuals represented in the frozen gene pool, as well as the number of samples from each individual, are dependent on availability, fertility, and many other considerations.

Frozen Semen Banks

Although the following discussion concentrates on semen banking, in many ways it applies to ova and embryo banks as well. The space required to store a frozen gene pool is generally small, and the cost for storage of the frozen material varies, but is relatively low compared to maintenance costs of live animals in wildlife refuges and zoological parks. The greatest expense in semen banking (even greater for ova and embryos) is the time, personnel, and facilities required to develop a suitable freezing technique for each species. Ideally, this basic research should be completed before the samples are stored; however, one should take advantage of every opportunity to collect semen from unstudied species (see Chapter 8).

Cryopreserved semen is a valuable resource for species propagation and gene pool preservation. The elimination of deleterious alleles in breeding populations may be accomplished by carefully selecting semen samples used in propagation programs, and genetic diversity may be enhanced by carefully planned artificial inseminations. In cases where gene pool rejuvenation is needed, semen samples will often be less expensive to transport, import, or export than living animals. Cryopreserved semen can be used to make more effective use of proven sires, to prolong the use of sires after death, to conduct matings out-of-season, and to avoid problems of disease transmission among living animals. Other, largely unexploited uses for cryopreserved gene pools include applications in forensic and medical science, environmental monitoring programs, and basic genetic and evolutionary research.

SECTION II

PROCEDURES FOR THE CRYOPRERVATION OF TISSUES

CHAPTER 6

STABILITY OF MACROMOLECULES DURING LONG TERM STORAGE

Herbert C. Dessauer and Robert A. Menzies

Preserving materials of biological origin in the vital state involves steps to minimize denaturing influences encountered when cells are removed from their natural environments. In general, this includes excluding light and other forms of radiation, minimizing exposure to microorganisms, maintaining low temperature, and taking steps to maintain the material as close to its *in vivo* chemical environment as possible. If the objective is to preserve the majority of cellular components as well as the biological activity and structures of proteins and nucleic acids, quick freezing with dry ice or liquid nitrogen is commonly used, especially under field conditions. Regardless of the type of material or its intended use, low temperature storage in liquid nitrogen or in an ultracold freezer is usually recommended for long term storage.

The greatest value of frozen tissues is for studies involving compounds whose structure does not depend on the integrity of cellular or organellar membranes. In fact, freezing, which slows both chemical and enzymatic modification, is often the only way to preserve the molecular integrity of compounds of interest. The structure of most compounds of low molecular weight, such as metabolites or environmental pollutants, remains unchanged during long term cold storage. However, precautions must be taken if the compounds of interest are prone to oxidation (e.g., certain lipids). If the rate of freezing is too slow and/or not done in a vacuum or under nitrogen, both chemical and enzymatic oxidation are likely (Fennema and Sung, 1980).

Knowledge of the stability of proteins during long term storage is surprisingly sparse. Sensabaugh and colleagues (1971a), in a short review of the subject, observed that denaturation rates of proteins vary widely; they concluded, however, that many proteins are far more stable than is generally assumed. For example, remnants of blood samples used in Nuttall's (1904) classical immunological study of mammalian relationships served as experimental material for one of the most significant tests of protein stability ever conducted. Keilen and Wang (1947), who carried out the study, estimated that 70 to 85% of the hemoglobin, carbonic anhydrase, catalase, glyoxylase and choline esterase remained active in these blood samples that had been kept in the dark under aseptic conditions at room temperature for approximately 42 years.

Some proteins retain activity for surprisingly long periods, even in tissues exposed to extremely adverse conditions. For example, of seventeen proteins that are commonly examined in electrophoretic studies, only alcohol dehydrogenase was undetectable in tissue samples from mammals 12 hours after death (Moore and Yates, 1983). Sensabaugh and colleagues (1971a,b) found that eight of 11 proteins in an 8-year-old sample of dried blood retained at least some activity. Plasma albumin and esterase activity were detectable in tissue samples taken from mammal skins stored for up to 16 years as standard museum preparations (J. L. Patton, pers. comm.). Albumin in muscle tissue from a mammoth frozen in northern Siberia for an estimated 40,000 years, and which had probably undergone numerous freeze-thaw cycles, maintained sufficient immunological specificity to demonstrate that the species is properly classified with the elephants (Lowenstein et al., 1981). Frozen, and even cooked, invertebrate tissues have been useful for forensic work (Krzynowed and Wiggen, 1981). Knowledge of the influence of environmental factors on the survival of biological activity is of major importance for scientists concerned with forensic problems (McWright et al., 1975; also see Chapter 4).

A critical question of special interest to managers and users of frozen tissue collections is how storage at ultracold temperatures affects the stability of proteins and nucleic acids. For the preservation of germ cells or collection of tissues for subsequent production of cell cultures, cells must be frozen at carefully controlled rates in the presence of cryoprotectants in appropriate concentrations (see Chapter 8). If these precautions are not taken, the production of intracellular ice and accompanying increase in concentrations of intra- and extracellular solutes may result in loss of cell viability. Reduction of cell viability is usually caused by damage to membranes rather than proteins, even though one would expect proteins to denature at subzero temperatures because of the interactions of hydrophobic groups (Mazur, 1970). However, several algal and liver cytochrome P450 mixed-function oxidases, which are notoriously labile membrane-bound enzymes, did not lose activity over several freeze-thaw cycles (see beyond). Unexpectedly, most soluble proteins appear to retain at least some activity indefinitely when preserved at ultracold temperatures.

Most nucleic acids are at least as stable as are proteins during long term cryopreservation. Generally, tRNA and rRNA can be isolated effectively from frozen tissues. On the other hand, certain types of studies are not possible using DNA extracted from frozen tissues. Intact DNA molecules are of importance to sequence studies (Sanger et al., 1977; Sanger, 1981). Because of the great length and rigidity of DNA chains in chromatin and DNA-protein complexes, the freeze-thaw process usually causes shearing and fragmentation. However, even partially sheared DNA preparations are valuable for many hybridization and gene cloning studies (Munjaal et al., 1981). Many workers are studying mitochondrial-DNA because it is easily isolated intact from purified mitochondria (Brown et al., 1979; Brown, 1980). In attempts to isolate either nuclear or mitochondrial-DNA, however, precautions must be taken to inhibit DNase activity as tissues are thawed. Denaturation of DNase with alcohol and/or inhibition with EDTA are two means of preserving DNA structure (Sibley and Ahlquist, 1981a).

Chemical Modifications and Storage Effects

Preservation in the frozen state for extended periods of time can lead to decreases in specific activity and minor changes in composition and conformation of certain proteins. These changes may be detected as alterations in the banding pattern of isozymes resolved electrophoretically, usually appearing as extra bands of activity that migrate at slightly slower or faster rates (Harris and Hopkinson, 1976).

Biochemical processes commonly resulting in changes in electrophoretic phenotypes include: (1) deamination of asparagine residues (Gracy, 1975); (2) oxidation of heme iron (Jimenez-Marin and Dessauer, 1973) or sulfhydryl groups of cysteine residues; (3) polymerizations traceable to sulfhydryl interchange reactions (Harris and Hopkinson, 1976); (4) proteolysis; (5) glycosylation (Bunn and Higgins, 1981); (6) sialylation (Parker and Bearn, 1960; Blumberg and Warren, 1961); (7) low levels of coenzymes and/or stabilizing ions in the chemical environment (Yoshida, 1966); and, (8) formation of conformational isomers (Kitto et al., 1966; Gockel and Lebherz, 1981). For example, deamination of proteins leads to the appearance of bands of faster mobility and to decreases in the concentration of the slower migrating, unaffected molecules. This appears to be a common modification that can occur both *in vivo* (Gracy, 1975) as well as *in vitro* (Sensebaugh, pers. comm.). Oxidation of ferrous iron in proteins such as hemoglobin and myoglobin results in the appearance of a band of slower mobility than that seen in unaffected proteins (Jimenez-Marin and Dessauer, 1973). The oxidation of sulfhydryl groups of cysteine residues may form disulfide (cystine) bridges between different proteins that leads to an increase in molecular weight. The "multihemoglobins" of turtles, which have weights of two, three, or four times unpolymerized hemoglobin, exemplify this effect (Svedberg, 1934). Vertebrate glucose-6-phosphate dehydrogenase disaggregates and loses activity in environments that lack NADP (Yoshida, 1966; Harris and Hopkinson, 1976). After muscle samples of birds are frozen and thawed, additional creatine kinase isozymes that appear to be conformational variants have been observed (Dawson et al., 1967).

Use of Frozen Tissue Collections for Stability Studies

Collections of frozen tissues have provided source material for examining the stability question. Transferrins in snake plasma frozen in 1953 could not be distinguished immunologically from those in fresh blood collected from the same species 16 years later (Mao and Dessauer, 1971). Protein phenotypes of a wide variety of enzymes were scoreable following electrophoresis of homogenates from tissues stored in frozen collections for eight or more years (Lawson and Dessauer, 1979). The repeated freezing and thawing of tissues, however, commonly results in a gradual loss of activity of certain enzymes. With tissues of lobsters, for example, isocitrate dehydrogenase, 6-phosphogluconate dehydrogenase, and leucyl-proline peptidase activities were lost rapidly. Enzyme stability, however, varies from species to species. For example, many of the same enzymes that were unstable in lobster tissues survived repeated freeze-thaw cycles in tissues from crocodiles, marine turtles, and several species of marine fishes stored in the same freezer at -20°C. Thermally labile variants of some proteins have been observed. Whereas the common lactate dehydrogenase of tail muscle of lobsters is stable, activity of a genetic variant was lost after only two freeze-thaw cycles (Raccuia and Menzies, pers. comm.). Even proteins in tissues with high proteolytic activity, such as venoms of viperid snakes, retained their activities for extended periods of time (Russell et al., 1960; Jimenez-Porras, 1961, 1964). Klebe (1975) described a simple electrophoretic method that should be useful in comparing the stability of proteins.

Investigators at the Evolutionary Genetics Laboratory of the Museum of Vertebrate Zoology, University of California at Berkeley, have undertaken several studies of protein stability. They found that whole carcasses and individual organs of mice stored frozen at -76°C for 3.5 years showed no electrophoretically detectable changes in the activity of alcohol, glycerol phosphate, lactate, malate, isocitrate, glucose-6-phosphate, xanthine, and sorbitol dehydrogenases; malate enzyme; superoxide dismutase; glutamate-oxalacetate transaminase; phosphoglucomutase; several esterases and peptidases, glucose phosphate and mannose phosphate isomerases; and albumin. Tissue homogenates were less stable than organs during prolonged storage, however the most common type of change observed in proteins in homogenates was a decrease in activity rather than a change in electrophoretic banding pattern (J. L. Patton, pers. comm.).

Reversing Detrimental Effects of Cryopreservation

The addition of reducing agents (e.g., dithiothreitol or mercaptoethanol), coenzymes, and activating ions to homogenates of previously frozen tissues stabilizes some enzymes and may reverse some of the adverse effects of long term storage (Harris and Hopkinson, 1976). Proteins are apt to denature when in dilute solution because of surface effects. To minimize this process, perturbants such as sucrose, mannitol, or glycerol are sometimes added to homogenizing fluids and diluents; "inert" bovine plasma albumin is used as a perturbant to stabilize the highly dilute antigen and antibody solutions used in microcomplement fixation (Champion et al., 1974).

By using specialized thawing procedures, it is possible to recover the tissue organization needed for histological and histochemical studies (see Chapter 7). Although the technology is not yet available to reverse the damage to cells caused by rapid freezing in the absence of a cryoprotectant (see Chapter 8), future advances in cryobiology may make it possible to culture cell lines from untreated tissues stored in the many frozen collections in the U.S.

CHAPTER 7

PROCEDURES FOR COLLECTING AND PRESERVING TISSUES FOR MOLECULAR STUDIES

Herbert C. Dessauer, Robert A. Menzies, and David E. Fairbrothers

Procedures for collecting and handling tissues for future molecular studies can be carried out, even in the field, by individuals with minimal training. Although each major group of organisms presents special problems, many procedures are common to all groups. This chapter outlines some techniques used to collect, transport, preserve, and handle tissues in the field and laboratory.

Tissue Collection

Animal Tissues.—Blood and other tissues should be sampled while the specimen is alive or as soon after its death as is feasible. As conditions are often less than favorable in the field, collectors should be aware of the importance of keeping their instruments, containers, and reagents clean, and of placing tissue samples in the cold and away from light as rapidly as possible. Tissues should be packaged in plastic cryotubes (e.g., "Nunc" tubes, Vangard International, Inc.), plastic bags (e.g., "Whirl-Pac" bags, American Hospital Supply Corp.), or wrapped tightly in aluminum foil, excluding as much air as possible. Small samples of blood can be collected in heparinized hematocrit or micro-tubes; larger samples are most efficiently collected from the heart or caudal vessels using a heparinized syringe. A doppler ultrasonic device (Brazaitis and Watanabe, 1982) is useful for locating externally the position of the heart. Gorzula and colleagues (1976) described a method for obtaining blood from caudal vessels; this technique has proven useful in work with larger animals such as crocodiles.

Blood cells should be separated from plasma prior to freezing. Commercial hand centrifuges or a lightweight, plastic centrifuge (Dessauer et al., 1983; Figure 7.1) are useful for separating blood cells from plasma under field conditions. Great care must be taken in labeling tubes and packages; pens containing ink that is resistant to moisture and ultracold temperatures are available commercially. Labels written with lead pencil on roughened surfaces are very stable, and those etched into plastic or glass tubes with a diamond-tipped pen are virtually permanent. The label should include the field number, tissue type, and species name (see Section III).

As soon as possible after collection and packaging, most tissues should be quick-frozen by dropping them directly into liquid nitrogen or by covering them with dry ice. However, quick-freezing generally shatters fragile hematocrit and micro-tubes filled with tissue fluids; such tubes must be frozen slowly before being subjected to ultracold temperatures. Household freezers are useful for this purpose, and are also

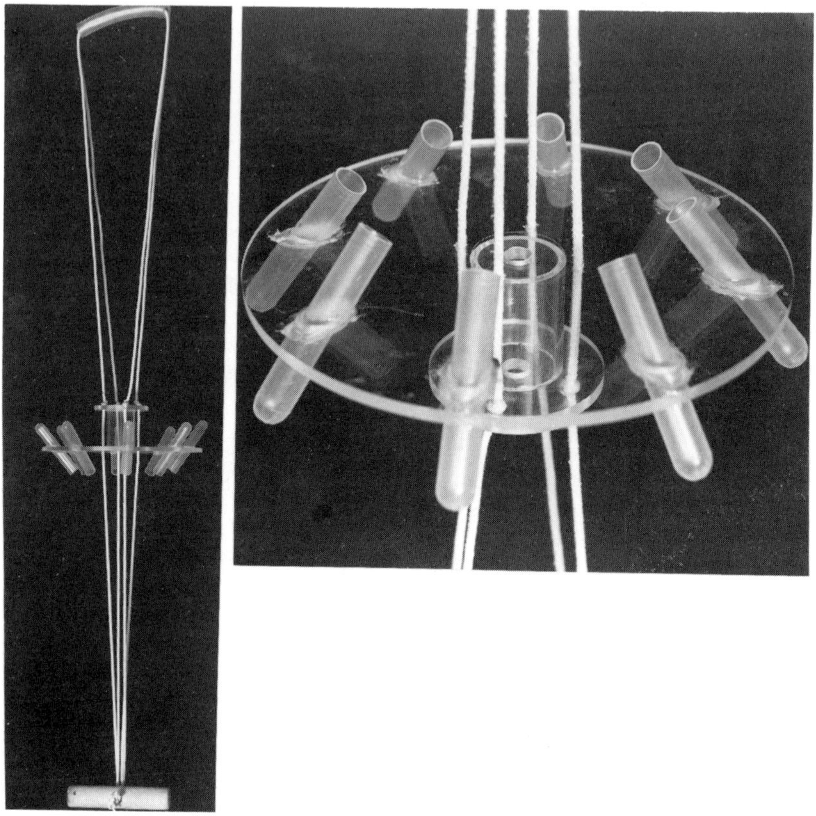

Figure 7.1. A simple, hand-held centrifuge for field use. The device consists of plastic tubes cemented in a plexiglas disk. The centrifuge operates on the "Chinese top" principle by alternate tension and relaxation of the cords (see Dessauer et al., 1983).

adequate for short term storage of all tissue samples (in emergencies, a salt-ice mixture will substitute as a temporary refrigerant). Fragile capillary tubes and microtubes can be inserted into the slots of corrugated cardboard for protection during long term storage.

Refrigeration is not required for tissues collected for certain purposes. Saturation with 75 to 80% ethanol is sufficient to preserve material for many DNA studies. The nucleated red blood cells of nonmammalian vertebrates are an especially convenient source of DNA because they lack tough connective tissue. Tissues other than red cells should be cut into pieces approximately 1 cm in diameter to allow the ethanol to penetrate them quickly. After immersion in alcohol for at least two days, the moist tissues may be transferred to plastic bags for storage or shipment (Sibley and Ahlquist, 1981a). Although not recommended as a long term storage procedure, immersion of tissues in an aqueous solution containing 2% 2-phenoxyethanol preserves proteins of blood plasma and certain other tissues for short periods of time. Plasma albumin and a number of the enzymes commonly examined in electrophoretic and immunological investigations retain their physical, catalytic, and immunological properties for at least three weeks in this preservative (Nakanishi et al., 1969). Lenses of vertebrate eyes, collected for use in sequence studies of alpha-crystallin, can be preserved in saturated guanidine hydrochloride (de Jong, pers. comm.).

Botanical Tissues.–Leaves, pollen, seeds, fern spores, and tubers of vascular plants have been preserved successfully for subsequent use in studies at the molecular level (Hunziker, 1969; Jensen and Fairbrothers, 1983). Seeds, pollen, and fern spores should be harvested only when mature. The collector should be aware that hot and dry weather prior to harvest may cause synthesis of "storage proteins" to cease prematurely.

The following protocol is recommended for the preservation of seeds from most plants: (1) remove fleshy portion of seed; (2) dry the seed; (3) place in vacuum-sealed container; and, (4) store in the dark at, or below, zero degrees C. Seeds of many taxa have remained viable for up to 10 years when stored under the above conditions. Pollen and fern spores may be treated in the same manner, however they should be screened to remove debris and treated with organic solvents to extract lipids. Macromolecules in pollen and fern spores stored at, or below, -30°C are known to be stable for at least four years. Pollen should not be ground until extracts are to be prepared. For the preservation of vegetative tissues, mainly leaves, the material should be washed in distilled water and rapidly frozen in liquid nitrogen for subsequent storage at ultracold termperatures. In some taxa, a "senescent" phenomenon occurs in which many proteins disappear from the leaves with seasonal aging. It is, therefore, important that the collector record the age of the leaves at harvest.

Transport of Frozen Tissues from Field to Laboratory

Frozen tissues are usually transported in either polyfoam boxes packed with dry ice or in liquid nitrogen containers. Airlines have special rules governing the shipment of these refrigerants (see Chapter 9). Dry ice containers are accepted as baggage once the appropriate form is submitted to the airline agent; however, liquid nitrogen containers, because they are easily spilled, present a greater problem. Assuming that the container is properly constructed and labeled, the decision as to whether or not it is accepted as baggage on a particular flight depends largely on the discretion of the airline agent and the pilot. The shipper should be well aware of all regulations concerning air transport of liquid nitrogen (see Chapter 9). As a last resort, the liquid nitrogen can be poured out of the container immediately before checking it as baggage. Most standard liquid nitrogen tanks are so well insulated that they will maintain a large mass of tissues in the frozen state for 10 or more hours, even in the absence of liquid nitrogen. The Union Carbide Corporation has recently introduced a small liquid nitrogen tank (3DS "Dry Shipper") that contains an absorbant which keeps liquid nitrogen from spilling during transport; perhaps this container will eliminate many of the problems associated with air transport of liquid nitrogen.

Dry Preservation of Tissues

Many proteins in carefully dried tissues are stable for short periods of time at room temperature and for much longer periods in household refrigerators or deepfreezers. Endocrinologists have long known that acetone powders, solids precipitated from tissues with cold acetone, retain many of the activities of the parent tissue. Botanists find that acetone powders of seeds and pollen also retain many biological activities. Plant tissues, however, are often difficult to preserve for later macromolecular studies because of the presence of phenols, polysaccharides, and lipids. Lipids should be extracted from seed meal (flour) before the meal is stored. The biological activity of certain leaf macromolecules is preserved, even after rapid desiccation at 45°C in a vacuum oven. After this treatment, these tissues can be stored in a standard freezer. Freeze-drying of seeds, pollen, and fern spores often leads to loss of biological activity; in contrast, many animal tissues, including plasma, are largely unaffected by this treatment. Mellor (1978, pers. comm.) suggests that vacuum drying of frozen tissues over a dessicant such as silica gel is less likely to damage sensitive proteins than is the usual method of freeze-drying. The activity of many freeze-dried proteins survives without refrigeration.

Use of Tissues for Histological Studies and Isolation of Organelles

In the past, frozen tissues have proven to be of little value for histological and cytological studies. If frozen, thawed, and processed in the usual manner, soft tissue structure is usually lost (but see Chapter 8). However, some microscopic structure remains in muscle and other tissues in which there is much associated connective tissue.

There are at least two approaches that may enable one to retrieve histological information from frozen tissues (J.M. Kerrigan, pers. comm.). First, it may be possible to simply excise a portion of the tissue and treat it as a frozen section. *In situ* enzyme histochemical analyses may be performed using these sections. In another approach, the excised sample is allowed to thaw very slowly while immersed in a fixative such as buffered 2% glutaraldehyde or buffered 10% formaldehyde. The addition of dimethylsulfoxide to the fixative solution might promote rapid penetration of the reagent and thus more effective fixation. This technique should be adequate for many applications in light microscopy and certain low magnification ultrastructure studies.

Recovery of organelles from frozen tissues is generally poor because organelle substructure is often destroyed in the freezing process. Organelles that are not membrane bound, such as ribosomes, are usually recovered in good yield. However, under normal freezing conditions, a large percentage of membrane-bound organelles, such as mitochondria and lysosomes, are usually ruptured. Nevertheless, some intact organelles can usually be recovered from frozen tissues rich in that particular organelle. For example, Landsman and colleagues (1981) obtained up to a 50% yield of mitochondria from frozen rodent liver. Although the yield of intact mitochondria will vary with species and tissue type, isolation techniques have been developed that should be of broad utility in work with frozen materials (Landsman et al., 1981; Munjaal et al., 1981; Southern, 1975).

CHAPTER 8

PROCEDURES FOR COLLECTING CELL LINES UNDER FIELD CONDITIONS

Robert J. Hay and George F. Gee

Cryopreservation of living cells requires special collecting, freezing, and storage procedures if the cells are to survive. Cell damage is most likely to occur during the freezing and thawing process. Some cells from most animals will survive freezing (-196°C) and thawing, but few cells will survive without a cryoprotectant such as glycerol or dimethylsulphoxide (DMSO). For every species, tissue, and freezing system there is an optimum cryoprotectant concentration and freezing rate (Mazur, 1980). The cryoprotectant must be concentrated enough to protect the cells from freeze damage, yet dilute enough to avoid chemical injury to cells. Ideally, the rate of freezing should be precisely controlled, with rate depending on variables such as species, size of sample, cryoprotectant used, and the system used to freeze the sample. Generally, too fast a cooling rate results in death due to formation of ice crystals within the cells; too slow a rate causes death from the chemical consequences of solute concentration. Obviously, the best combination, one designed for a specific tissue, freezing system, and species, will result in the maximum number of live cells recovered from the sample. However, it is possible to store and recover viable animal cells without a highly specialized program.

Often a system that works for one animal is successful with the same tissue in another animal. For example, Seager and colleagues (1978) and Graham and colleagues (1978) have used domestic animal semen preservation techniques to preserve semen from hundreds of nondomestic species. They, and others, have recovered live sperm from most frozen semen samples, and through artificial insemination, have obtained young in many cases. Sperm from a species of nondomestic bird, frozen using a technique developed for domestic fowl, has been used to produce viable young (Gee and Sexton, 1979).

Both semen samples and tissue biopsies can be easily obtained under field conditions without permanent injury to the donor animal. The equipment and supplies needed to establish proper freezing conditions in the field are not elaborate: alcohol, freezing medium, and liquid nitrogen in an appropriate tank (see Chapter 7 and Maure, 1978). If the freeze rates used are less than optimal, manipulation during the thawing process may increase the recovery of live cells (Mazur, 1980).

Two general protocols for tissue biopsy and preservation recommended by the American Type Culture Collection are described for consideration by field collectors:

1. For optimum recovery, the biopsy (5-10 mm in diameter) should be collected aseptically and placed in a vial or flask filled with tissue culture medium containing 10% serum and antibiotics (penicillin, streptomycin,

gentamycin, and amphoteracin B). Skin biopsies are collected using sterile instruments after thorough wiping of the tissue surface with 70% ethanol or other antiseptic. If the culture is viable, it is processed, usually within 3 to 4 days, using standard procedures (Hay et al., 1982). The cell culture can be studied immediately, or frozen for long term storage (Hay, 1978).

2. If the above procedure is not practical, an alternative method known to be satisfactory for human skin biopsies may be used. The tissue sample is collected and treated aseptically as above. It is then placed in a tissue culture medium containing 10% serum, antibiotics, and DMSO and minced using sterile lens scissors to yield fragments of about 1 mm in diameter. We recommend the use of 10 to 12% DMSO, but other concentrations have been used successfully (Taylor et al., 1978). If possible, one should allow the tissue to equilibrate with this "freeze medium" for 2 to 3 hours at 4°C, after which its temperature should be lowered to -50°C at a rate of about one degree per minute. The frozen sample can then be placed directly into liquid nitrogen. Gradual cooling can be accomplished using insulated containers or other devices available commercially (Hay, 1978). Otherwise, successful cultures have been established using tissue samples placed in the appropriate medium and frozen immediately in liquid nitrogen (Taylor et al., 1978; R. J. Baker, pers. comm.). It may also be possible to recover cells treated in the above manner and frozen in dry ice. The time of storage at dry ice temperatures should be minimized, however, and storage in liquid nitrogen is definitely preferred.

CHAPTER 9

REGULATIONS GOVERNING ACQUISITION AND TRANSPORT OF FROZEN TISSUES

Importation of Frozen Tissues

The scientist intending to import frozen animal tissues (or other animal parts) must adhere to all applicable U.S. Federal Wildlife Regulations. In general, tissues and tissue components are subject to the same regulations imposed on the importation of standard museum specimens (skins, skeletons, and fluid specimens). U.S. Department of the Interior Form 3-177, "Declaration for Importation of Fish or Wildlife" (Figure 9.1) must be submitted to customs agents at the port of entry. Persons planning to import living animals, semen, or fertile eggs must apply for a permit from the U.S. Department of Agriculture Animal and Plant Health Inspection Service using Veterinary Services Form 17-129 for each shipment (Figure 9.2). U.S.D.A. Publication 9 CFR Part 92 describes in great detail regulations on the importation of animals and animal parts. The Center for Biosystematics Resources of the Association of Systematics Collections (Museum of Natural History, University of Kansas, Lawrence, KS 66045) will provide interested persons with up-to-date information on U.S. Federal Wildlife Regulations, federally controlled species, and state protected species.

Because of the potential threat to domestic poultry, the importation of frozen avian tissues is subject to additional federal regulations. Avian tissues fall under the category of "controlled materials, organisms, or vectors" as defined by the U.S. Department of Agriculture. Any facility wishing to import avian material into the U.S. (or transport avian tissues within the U.S.) must first obtain a permit through the U.S.D.A. Animal and Plant Health Inspection Service (see Figure 9.3, VS FORM 16-3). The U.S.D.A. Parent Committee on Foreign Pathogens and Vectors will evaluate the application and recommend approval or disapproval. In the past, this committee has required that the laboratory be inspected and found biologically secure, that all bird tissues be processed under a vertical flow biological hood, that all personnel involved in the project not have contact with live birds or poultry for 60 days after working with the imported materials, and that all bird tissues be confined to the approved processing area and storage freezers. Thus, transport (loaning) of viable avian materials from an approved to an unapproved laboratory is prohibited.

Air Transport of Dry Ice and Liquid Nitrogen

Tissues transported by air are generally shipped on dry ice or in liquid nitrogen. Both of these susbtances are classified as Restricted Articles under the International Air Transport Association *Dangerous Goods Regulations*, 24th edition, effective 31 December 1982. Dry ice has been designated as Hazard Class ORM-A, and must be so marked. A "Shipper's Certification for Restricted Articles" (supplied by indivi-

USFWS Form 3-177
(revised Sept 1978)

Form Approved O M B No 42-R1476

U.S. FISH AND WILDLIFE SERVICE

DECLARATION FOR IMPORTATION OR EXPORTATION OF FISH OR WILDLIFE

Indicate One:
☐ import ☐ export

Port of:

Date:

For Imports Only

Customs Identification No.:

Name of Carrier:

Airway Bill or Bill of Lading No.:

Imported or Exported via:
☐ air cargo ☐ ocean cargo ☐ truck
☐ rail ☐ mail ☐ personal baggage
☐ automobile: license no. ________
state ________

Location where wildlife is available for inspection:

Package or Bale Marks and Nos.:

(indicate one)

Please Type or Print Legibly

☐ U.S. Importer of Record

☐ U.S. Exporter ________________ (name) ________________ (address - street, city, state, zip code)

Foreign Consignor or Consignee ________________ (name) ________________ (address - street, city, country)

Customs Broker or Agent ________________ Shipping Agent or Freight Forwarder ________________

Furnish All Information Below (invoices or lists providing required information may be attached)

QUANTITY	SCIENTIFIC NAME Genus Species	COMMON NAME	DESCRIPTION If live, so state. If product, describe.	DECLARED VALUE	COUNTRY OF ORIGIN

U.S. License and/or Permit Nos.:

Foreign License and/or Permits: Country No.

I certify the information contained herein is true and complete to the best of my knowledge and belief.

________ signature ________ date

Action Taken: Date:
☐ cleared
☐ clearance refused
☐ seized

________ % of Wildlife Inspected:

Officer Signature and Badge No.:

FOR OFFICIAL USE ONLY

SEE REVERSE OF THIS FORM FOR INSTRUCTIONS AND PRIVACY ACT NOTICE

Figure 9.1. U.S. Department of the Interior Form 3-177. All imported animal tissues must be declared on this form at port of entry.

No animals, semen, poultry or hatching eggs may be imported unless a completed application has been received (9 CFR 92).

U.S. DEPARTMENT OF AGRICULTURE
ANIMAL AND PLANT HEALTH INSPECTION SERVICE
VETERINARY SERVICES

FORM APPROVED
BUDGET BUREAU NO. 40-R2981

APPLICATION FOR IMPORT OR IN TRANSIT PERMIT

(Animals, Animal Semen, Poultry, Birds and Hatching Eggs)

RCS# 34-VS-33

INSTRUCTIONS TO IMPORTER: Complete and submit one copy to the Veterinary Services, APHIS, U.S. Department of Agriculture, Federal Center Building, Hyattsville, Maryland 20782. Prepare a separate application for each shipment.

1. NAME AND ADDRESS OF SHIPPER IN COUNTRY OF ORIGIN

2. NAME AND ADDRESS OF IMPORTER *(Include Zip code)*

TELEPHONE NUMBER *(Include Area Code)*

3. PORT OF EMBARKATION *(From Canada show only for ocean vessel or airplane shipments)*

4. COUNTRY FROM WHICH SHIPPED

5. MODE OF TRANSPORTATION *(Name of Airline or Vessel, flight no.)*

6. ANIMALS, ANIMAL SEMEN, POULTRY, OR HATCHING EGGS

NO. (A)	BREED (B)	SPECIES (C)	DESCRIPTION *(Sex, Age, Registered Name and No., Tattoo, Tag No., other Markings)* (D)

6E. PURPOSE OF IMPORTATION

7. ROUTE OF TRAVEL INCLUDING ALL CARRIER STOPS ENROUTE *(From Canada show route of travel **only** for ocean vessel or airplane shipment)*

8. PROPOSED SHIPPING DATE *(From Canada show only for ocean vessel or airplane shipment)*

9. PROPOSED ARRIVAL DATE

10. UNITED STATES PORT OF ENTRY

11. NAME AND MAILING ADDRESS OF PERSON TO WHOM DELIVERY WILL BE MADE *(After quarantine, when required) (Include Zip code)*

TELEPHONE NUMBER *(Include Area Code)*

12. WHERE DELIVERY WILL BE MADE IN U.S. *(After quarantine, when required) (Location of place)*

13. REMARKS

14. SIGNATURE OF IMPORTER

15. DATE SIGNED

VS FORM 17-129
JAN. 1978

PREVIOUS EDITIONS MAY BE USED

Figure 9.2. U.S. Department of Agriculture Form 17-129. A U.S.D.A. permit must be obtained for each shipment of imported living animals, animal semen, or fertile eggs.

No controlled material, organisms or vectors may be imported or moved interstate unless the data requested on this form is furnished and certified. *(9CFR 94, 95 and 122)*

U. S. DEPARTMENT OF AGRICULTURE
ANIMAL AND PLANT HEALTH INSPECTION SERVICE
VETERINARY SERVICES
FEDERAL CENTER BUILDING
HYATTSVILLE, MARYLAND 20782

APPLICATION FOR PERMIT TO:

☐ **IMPORT CONTROLLED MATERIAL**

☐ **IMPORT OR TRANSPORT ORGANISMS OR VECTORS**

FORM APPROVED: OMB NO. 40-R2717

1. MODE OF TRANSPORTATION

2. U.S. PORT(S) OF ENTRY

INSTRUCTIONS: Submit 2 copies to address above. Attach additional sheets, if necessary.

3. **TO:** *(Name, address, and phone no. of applicant - include Zip Code)*

4. **FROM:** *(Name and address of shipper - include Zip Code)*

5. DESCRIPTION OF MATERIAL *(Name of material, country of origin, animal source, etc.)*

6. QUANTITY OF MATERIAL TO BE IMPORTED AND FREQUENCY OF IMPORTATIONS

7. PROPOSED USE OF MATERIAL, EXPECTED COMPLETION DATE, AND FINAL DISPOSITION TO BE MADE

8. DESCRIPTION OF APPLICANT'S FACILITIES AND EQUIPMENT FOR HANDLING MATERIAL

9. QUALIFICATIONS OF TECHNICAL PERSONNEL WHO WILL BE WORKING WITH THIS MATERIAL

10. METHOD OF TREATMENT OF MATERIAL *(Disease Safeguard)*

11. WORK OBJECTIVES, PROPOSED PLAN OR WORK AND ADDITIONAL PERTINENT INFORMATION

12. PERTINENT PUBLISHED PAPER OR ASSTRACT *(Please attach copy, if available)*

CHECK IF COPY IS ATTACHED ☐

I certify this material will be used in accordance with all restrictions and precautions as may be specified in the permit.

13. SIGNATURE OF APPLICANT

14. TYPED NAME OF OFFICIAL SIGNING

15. DATE SIGNED

16. TYPED TITLE OF OFFICIAL SIGNING

VS FORM 16-3
AUG. 1976

REPLACES VS FORM 16-3 (3/73) AND VS FORM 16-27 (7/74), WHICH ARE OBSOLETE.

Figure 9.3. U.S. Department of Agriculture Form 16-3. A U.S.D.A. permit is required for importation and interstate transport of controlled materials, organisms, and vectors. Importation of frozen avian tissues and transfer of these tissues between U.S. collections requires a U.S.D.A. permit.

dual airlines) must be attached to the shipment. No more than 440 pounds of dry ice may be shipped in a single package.

Liquid nitrogen carried in non-pressurized, metal dewar flasks is authorized for air shipment, but only with the approval of the airplane's pilot. No more than 13 gallons per flask can be carried on a passenger carrying aircraft (98 gallons on a cargo aircraft). The flask must be marked "Nitrogen (*Liquid, Non-Pressurized*)," and must be designed or packaged to strongly discourage loading or handling in any position other than upright. The upright position of the container must be indicated prominently by arrows and the wording "KEEP UPRIGHT" placed at 120 degree intervals around the container. It must be prominently marked "DO NOT DROP" (see International Air Transport Association Restricted Articles Circular No. 6-D, section 3, part X, item 2001).

Problems With Federal Regulations

The complexity of federal regulations governing the importation of frozen tissues often leads to considerable confusion on the part of researchers and lengthy time delays in obtaining permits. Although the scientific community appreciates the spirit of these regulations, the multiplicity of forms and permits often has an excessively inhibitory effect on scientific, medical, and veterinary research. Much of the problem here stems from the fact that, until recently, few federal agencies were confronted with requests to import frozen tissues, and their initial reaction has been to deny such requests. In many cases, the scientist is required to describe in detail the nature of frozen tissues, their use, storage, and disposal. Although this information is important for the permit-granting process, the Workshop Panel members agreed that the present system involves much duplication of effort on the part of both scientists and agency officials.

The potential health and economic hazards associated with the importation by scientists of frozen animal tissues are miniscule. The tissues that are imported by most scientists (heart and skeletal muscle, liver, kidney, and blood; see Chapter 12) are rarely infected with pathogens or vectors. The very nature of the research performed using these tissues requires that they be maintained at ultracold temperatures to arrest biological activity (see Chapter 6). The tissue samples are generally very small and are handled with extreme care by the investigator (see Section III); most are never directly handled. The vast majority of people working with frozen tissue collections are geneticists and evolutionary biologists (see Chapter 11), rather than virologists or epidemiologists. Clearly, the purposeful isolation of pathogens and vectors should be carefully regulated. However, the importation and use of tissues for genetic and evolutionary studies does not warrant excessive regulatory measures.

The U.S.D.A's concern that the importation of frozen avian tissues may cause an outbreak of Newcastle disease in domestic poultry is a clear example of the general lack of communication between scientists and federal officials. Biologists are well aware that millions of birds of many species and from many geographic regions fly across the borders of the U.S. each year. Clearly, these live animals pose a far greater threat to the poultry industry than do a few hunded samples of frozen tissues collected from wild-caught birds outside the U.S. Moreover, unlike free-living birds that often come into direct contact with domestic poultry, the tissue samples remain stored in deepfreezers in the laboratory.

It is ironic that U.S.D.A. policy prohibits the importation of untreated ruminant tissues from countries with foot-and-mouth disease, but permits the direct importation of living animals from these countries provided they pass quarantine. In a recent case (O. A. Ryder, pers. comm.), an application to import a sterile skin biopsy of the endangered Mesopotamian fallow deer was denied by the U.S.D.A. This biopsy was to have been propagated for chromosomal analysis, providing information that may have proven useful in future attempts to propagate this species. This research was to have taken place in a laboratory that routinely handles tissue cultures from a wide variety of mammals. By their very nature, tissue cultures must be maintained in sterile, uncontaminated environments. The likelihood of an accidental introduction of foot-and-mouth disease to livestock in the U.S. from this material seems remote. Because the scientist requesting the biopsy would have followed routine procedures in handling and disinfecting the tissue culture materials, he could only have introduced the disease into domestic herds if the original biopsy had been contaminated, if the biopsy grew, and if the cells were then deliberately introduced to susceptile domestic livestock.

Conclusions

The Workshop Panel members, and most scientists, agree that federal regulations concerning the importation of biological materials are needed to protect domestic crops and livestock, safeguard the general public health, and to control illicit trafficking in endangered and threatened species. However, these regulations, though necessary, should not be self-defeating by inhibiting *bona fide* medical, veterinary, and basic research. Scientists who import frozen tissues for genetic and evolutionary studies now constitute a relatively large portion of the research community, and their use of these tissues should not be considered a threat to the health and welfare of the country. The Workshop Panel recommends, therefore, that the regulatory agency concerned, the United States Department of Agriculture, reevaluate its policies in this area and implement swift, constructive measures to facilitate the importation of frozen tissues for scientific research.

CHAPTER 10

SOURCES OF LIQUID NITROGEN AND DRY ICE IN FOREIGN COUNTRIES

Scientists conducting field work in foreign countries often find it difficult to locate sources of liquid nitrogen and/or dry ice to preserve tissue samples. The frozen tissue collection managers surveyed for this report were kind enough to provide us with addresses of such sources in countries they have visited. The resulting list of addresses, although far from complete, may be of assistance to anyone planning field work outside of the United States. The editors would be pleased to receive information on liquid nitrogen/dry ice sources in other countries.

For investigators planning to collect frozen tissues in AFRICA or ASIA, it is advisable to contact local university personnel or airline officials to obtain addresses of local liquid nitrogen or dry ice suppliers. For liquid nitrogen in EUROPE, one should contact Union Carbide Europe, Polarstrean Department, 5 rue Pedro-Meylan, 1211 Geneva 17, Switzerland to obtain a booklet titled "European Filling Station Network."

AUSTRALIA

CIG (Colonial Industrial Gases)--Outlets in major cities.

BOLIVIA

Cochabamba: Gasona Ltd., Calle Kolla, Cochabamba, Bolivia.

La Paz: Liquid Carbonic, Rio Seco, La Paz, Bolivia.

CANADA

Calgary: University of Calgary (Chemical Storeroom), Calgary, Alberta, Canada.

CHINA

Taiwan: Ali Company, Taipei, Taiwan, Republic of China.

Taiwan: San Fu Chemical Company, Ltd., Chu-pei Plant, No. 2-1, Tai Ho Road, Tai Ho Tsun, Chu-Pei, Hsin-Chu Hsien, Republic of China.

COSTA RICA

San Jose: Bodegas Gases Industriales, S.A., Ltd., Calle 6, Avenida 16, San Jose, Costa Rica.

FINLAND

Espoo: Oy AGA Ab, Karapellontie 2, Espoo 21, Finland.

Jyvaskyla: Kuljetusliike A&P Vuorio, Hakkutie 4, Jyvaskyla 45, Finland.

Oulu: Oy AGA Ab, Kasarmintie 22, Oulu 61, Finland.

Tampere-Nekala: Oy AGA Ab, Kuokkamaantie 12, Tampere-Nekala 41, Finland.

Turku: Oy AGA Ab, Akselintie 16, Turku 31, Finland.

Vaasa: Oy AGA Ab, Poikkikuja 4-6, Vaasa 34, Finland.

ISRAEL

Holon: (Dry ice) Tzinada Factory, New Industrial Area, Holon, Israel.

MEXICO

Mexico City: Infra del Centro, Via Gustavo Baz No. 56, Barrientos Tlalnepantla, Edo. de Mexico, Mexico.

PANAMA

Panama City: Gorgas Memorial Laboratory, Avenida Justo Arosamena, Panama City, Panama.

PERU

Cachimayo: Industrial Cachimayo, Cachimayo (approximately 10 km from Cusco), Peru.

Chiclayo: Oxigeno Chiclayo, SA, Colon No. 120, Chiclayo, Peru.

Lima: Liquid Carbonic del Peru, S.A., 2597 Avenida Venezuela, Lima, Peru.

Lima: Sociedad Quimica Industrial, Avenida Guillermo Dansy 846, Lima, Peru.

SOUTH AFRICA

Durban: AFROX Ltd., 95 Maydon Road, Box 1522, Durban, South Africa.

SWITZERLAND

Basel: CARBAGAZ, Kohlenstrasse 40, 4000 Basel 13, Switzerland.

Bern-Liebefeld: CARBAGAZ, Waldeggstrasse 38, 3097 Bern-Liebefeld, Switzerland.

Geneve: CARBAGAZ, Bd St-Georges 72, 1211 Geneve 8, Switzerland.

Lausanne: CARBAGAZ, Chemin du Grand Pre 4, 1000 Lausanne 16, Switzerland.

Zurich: CARBAGAZ, Forrlibuckstrasse 30, 8005 Zurich, Switzerland.

WEST GERMANY

Hollriegelskreuth: Linde A.G., Werkgruppe technische Gase, Dr. Karl von Linde-Strasse 6-14, 8023 Hollriegelskreuth, Federal Republic of Germany.

ZIMBABWE

Gweru: Oxyco Company, Bristol Road, Gweru, Zimbabwe.

Harare: Oxyco Company, Glasgow Road Industrial Site, Harare, Zimbabwe.

SECTION III

CURATION OF COLLECTIONS OF FROZEN TISSUES

CURATORIAL PROBLEMS UNIQUE TO FROZEN TISSUE COLLECTIONS

Robert J. Baker and Mark S. Hafner

In view of the fact that 95% of the collection managers surveyed maintain primarily frozen materials in their collections, and these collections are composed primarily of vertebrate tissues and blood (see Section IV), it is appropriate that we restrict our comments here to these kinds of collections. Managers of other kinds of collections (e.g., preserved cell lines and frozen semen collections) have, in the past, faced unique problems of their own; solutions to many of these problems have been found, largely because of widespread commercial interest in these collections.

Frozen tissue collections present not only the usual array of problems associated with the establishment and maintenance of systematic collections, but also several unique problems as well. Following are some general observations, comments, and recommendations, resulting from our efforts to establish organized and flexible frozen tissue curatorial programs at Texas Tech University (Baker) and Louisiana State University (Hafner). We are grateful for the knowledgeable advice provided us by James L. Patton, manager of the frozen tissue collection at the Museum of Vertebrate Zoology, University of California, Berkeley.

Cataloguing in the Field

In traditional systematic collections, specimens are collected in the field, assigned a collector's field number, brought back to the museum and catalogued. However, with regard to frozen tissues collected in the field, it is difficult to write a museum catalogue number on a frozen vial once it is returned to the laboratory because of the rapid accumulation of moisture on the outside of the vial. Therefore, we recommend establishment of a tissue collection catalogue (separate from the main collection catalogue) so that frozen tissues can be handled in a different manner than are traditional museum specimens. We use two, very different cataloguing procedures in our respective collections, both of which we feel are adequate. The Texas Tech collection maintains a series of hard-bound catalogues (see Figure III.1), each containing a unique sequence of catalogue numbers. Each collector or collecting party takes a bound catalogue into the field, and a tissue collection number is assigned to each specimen and written on the dry cryotube (using lead pencil, cryo-proof ink, or diamond-tipped pen) before the tube is placed in liquid nitrogen. Each tube is also labeled as to tissue type and genus (and species, if known) of the specimen from which the tissues were taken (see Chapter 7). On return to the museum, the tissue catalogue is immediately cross-referenced with the main collection catalogue.

Karyotyped by R. J. Baker Date 11 March 84 **TK 26076**
Collector Baker et al. Date 11 March 1984
Preparator KN Number 109
Museum TTU Number 40265

Freezer location: Box AA, Ultracold II

Sex: ♀ Species: Onychomys leucogaster

State: TX County: Winkler

Specific locale: 17 mi by Hwy 115 NE of Kermit, Rd to YT Ranch
S. B. Wright Jr.

Mitotic ✓ Meiotic NO Tissue cultured NO

Liver ✓ Kidney ✓ Heart ✓ Blood NO Muscle NO

Other

Coordinates	2N	No. Biarm	No. Acro	Photo No.	Comments
A1 10.6 X 104.4	48	48	0	—	
A2 12.5 X 102.5	48	48	0	—	
A3 15.3 X 116.4	48	48	0	7859	Standard
C4 18.9 X 106.4	48	48		7860	G-bands
5 X					
6 X					
7 X					
8 X					
9 X					
10 X					

Remarks:
Not yeast stressed – 27 min in hypo.

Figure III.1. Sample page of tissue collection catalogue used at Texas Tech University. Although this form is oriented toward karyological work, a modified version would be appropriate for most tissue collections.

In the mammal frozen tissue collection at Louisiana State University, each cryotube is labeled in the field with the collector's initials, the collector's unique field number, tissue type, and the scientific name of the tissue source. On return to the museum, the specimen is catalogued and assigned a unique frozen tissue collection number. This number, however, is not written on the cryotube, it merely serves as a convenient way to record the number of specimens in the collection. The critical marking on each tube is the collector's initials and field number (e.g., "MSH 1350"), which is used to retrieve specimens from the ultracold freezers. In the freezers, all specimens of a given taxon are stored together (see beyond); thus, the possibility of two different collectors having the same initials and field number for two specimens of the same taxon is extremely remote. This cataloguing procedure does not require the collector to take a bound tissue catalogue into the field and make entries into two catalogues (personal catalogue and tissue catalogue). Further, the collector who captures many more specimens than planned does not run the risk of using up all tissue numbers assigned to the particular catalogue volume. In both collections, we have found the practice of writing the scientific name of the animal on the cryotube to be extremely useful in subsequent curatorial and research work. In addition, we believe that it is wise to write the catalogue number (Texas Tech) or collector's number (LSU) twice on each tube to avoid the possibilty of the number rubbing off due to abrasion with nearby tubes in the liquid nitrogen tank.

Certain collection managers avoid the problem of relabeling wet tubes by transferring samples from field (plastic) cryotubes to cork-stoppered glass vials before placing them in the collection. We recommend against this procedure because cork stoppers are often not airtight, and transfer between vials usually requires thawing of the sample; both of these represent potential threats to the longterm stability of the tissues (see Chapter 6). We recommend the use of 2cc or 4cc plastic cryotubes with airtight gaskets (see Chapter 7) for both field collecting work and long term storage in the collection. Flat-bottom, plastic cryotubes are more expensive than glass vials, but the expense is minimal compared to the usual costs involved in collecting and storing frozen tissues. Cryotubes, unlike cork-stoppered vials, can be sterilized and reused once the sample is used or discarded. The survey revealed that most collections do, in fact, store tissues and blood in plastic cryotubes.

Storage in the Museum

Once samples have been catalogued in the field and returned to the museum, other problems in handling and storage are encountered. Thawing of samples should be avoided at all times during handling (see Chapter 6). Samples may be stored in an electrical freezer, on dry ice, or in liquid nitrogen. If freezers are used, they should be of the "ultracold" type so as to maintain samples at -70°to -90°C. (Most collections surveyed store their materials in ultracold freezers). These freezers may be either chest type or upright models; the former maintains more constant temperatures during use and is, therefore, less prone to mechanical failure over time. The upright models use less floor space, and freezer boxes are more easily retrieved from freezers of this design. Whichever model is selected, the freezer should be equipped with an alarm system that will sound in the event of electrical or mechanical failure. Most freezers are sold with a battery powered local alarm; at LSU we are installing a remote alarm system connected to the campus police station. With this system, our collection will be monitored around the clock, 365 days per year. Some form of back-up storage system (other freezers, liquid nitrogen, or dry ice) should be readily available in the event of freezer failure (90% of U.S. collections are so protected). Ideally, liquid nitrogen is better for longterm storage because of its much colder temperature (-196°C). However, if large numbers of samples are stored in the collection, it may be difficult to retrieve specific samples from the liquid nitrogen tanks, and continual replenishment of evaporated liquid nitrogen may become costly. On a cost per sample basis, an ultracold freezer provides the most convenient and efficient method for longterm storage of large numbers of samples.

Another curatorial problem unique to frozen tissue collections involves the organization of samples within the freezer or liquid nitrogen tank for ease of access. In freezers, moisture-proof boxes labeled with numbers (or letters) can be used to sort samples by locality, or preferably, taxon. A listing of the holdings in each freezer, complete with box number, contents of the box, and location in the freezer is maintained and routinely updated as samples are moved, used, loaned or discarded. We recommend against storing the entire collection in numerical sequence because most access needs will be by taxon, and added freezer costs (due to loss of cold) are involved when searching for specimens scattered throughout the freezer(s). We recommend that freezers be opened as rarely as possible; one freezer should be set aside as a "working freezer" for storage of tissue samples that are currently being studied.

A small number of samples stored in liquid nitrogen tanks is relatively easy to organize for efficient retrieval. Each tank will hold six canisters, and each canister will hold approximatively 30 cryotubes. Each tank should be assigned a letter, and each canister a number. A file should be maintained indicating the contents of each canister. Again, it is possible to organize samples by locality, taxon, or any other category desired. If liquid nitrogen tanks are used, however, many tanks are needed if the volume of samples handled is high. Liquid nitrogen refrigerators are available that will efficiently organize up to 15,000 2cc cryotubes with easy access to any tube. However, a primary problem is initial cost, and such units have a high daily loss of nitrogen and hence require considerable yearly expenditures to maintain.

Acquisition Policies

The acquisition policies of a frozen tissue collection will vary with its freezer storage space, level of curatorial support, research interests of the curator (and other associated investigators), and ultimately, the collection's level of funding. Because maintenance of frozen tissues is relatively costly, we recommend that the curator pay very close attention to what is actually being catalogued into the collection. We recommend against cataloguing long series of samples from easily obtained organisms unless there is an immediate plan to utilize such samples in a research program. Unfortunately, the current level of support of frozen tissue collections prohibits the luxury of cataloguing and maintaining all samples received.

Managers of frozen tissue collections should seize every opportunity to acquire tissue samples of rare, unusual, and exotic species. When possible, samples of threatened or endangered species should be sought, but these (and all samples) should be acquired legally and used judiciously. Acquisition of tissues from certain kinds of organisms requires special permits (see Chapter 9). Collection files should contain copies of collecting permits issued to the original collector of the specimens, and if the material is imported into the United States, the files should contain copies of the required U.S. Department of Agriculture importation form 3-177 (see Chapter 9, Figure 9.1). All samples deposited in the collection should be documented as thoroughly as possible, much in the same manner as traditional museum specimens (Lee et al., 1982). Except in rare cases, all samples should be represented by a traditional voucher specimen (skin, skull, or fluid specimen) deposited in an accredited systematic collection.

Deacquisition Policies

Unlike materials in conventional systematic collections, materials stored in frozen tissue collections are usually consumed as they are analyzed. Samples analyzed in museum laboratories and materials sent out on loan are rarely returned to the collection. The word "loan" therefore should be replaced with "gift" or "donation", when discussing transfer of frozen materials between collections and researchers. Because of this fact, much of the catalogued collection is only quasi-permanent, and a premium is placed on judicious dissemination of materials and efficient and flexible inventory procedures to keep track of deacquisitioned materials (discussed beyond).

We agree with the majority of survey respondents whose policy is to "loan" specimens only to carefully selected scientists (see Chapter 11). Selection should be based on the rarity of the specimen (and/or size of the sample) and the research direction of the potential recipient. Because samples are usually destroyed by the recipient, denials of requests for materials may be fairly common. It is the joint responsibility of the donor and recipient to be sure that transfer of the specimens is legal (i.e., the recipient may require special permits to handle the material; see Chapter 9). In every case, the naturalist who originally collected the tissues should be suitably acknowledged in any publication resulting from their use.

Shipments of frozen tissues are costly to package and transport. Our survey indicates that, at present, the donor (collection) pays tissue shipping costs 57% of the time and the recipient 43% of the time. In view of the fact that shipment is generally in one direction only (donor to recipient), we recommend that tissue shipment costs normally be paid by the recipient. *Investigators planning to obtain samples from frozen tissue collections for use in their research should provide for these costs in grant budgets.*

Computerized Inventory Systems

Ultracold storage space is very expensive to purchase and maintain, and it is important that materials be stored in a very space-efficient manner. It is therefore imperative that the access and inventory procedures for frozen tissue collections be extremely well organized. Ultracold freezers are very sensitive to even brief periods of temperature warm-up, and every second that a freezer door is open while a technician searches for a particular sample is energy consuming and could eventually contribute to freezer failure. In short, the curator (or technician) must know *exactly* where each sample is located *before* opening the freezer. Add to this the potential problem of wasted searches for samples that have been recently deacquisitioned (only 37% of the curators surveyed routinely update their catalogues), and the need for a flexible inventory system becomes even clearer.

Many of the unique curatorial problems posed by frozen tissue collections can be minimized through the use of a computerized inventory system. As a sample is "loaned" (= deacquisitioned), this deletion from the collection must be promptly recorded to avoid pointless and costly future searches for this sample. In addition, frozen tissue collection catalogues should be continually updated as specific tissues are homogenized or consumed, and as storage positions in the freezers are changed.

Important biochemical data, such as unique or marker alleles and the titer of antisera, should be recorded as they are obtained. A large number of data base management programs are currently available, most of which are adaptable for collection computerization.

SECTION IV

CHARACTERIZATION OF EXISTING COLLECTIONS

CHAPTER 11

SUMMARY OF FINDINGS FROM THE INTERNATIONAL SURVEY

Mark S. Hafner

In order for the Workshop Panel to deal effectively with the issue of frozen tissue collection management, it was first necessary to survey existing collections and query collection managers as to their collection holdings, uses of these holdings, sources of financial support, and plans for the future of their collections. In addition, the survey provided an opportunity to poll collection managers concerning a possible "national plan" for collections of viable tissues.

The survey consisted of a four page questionnaire which was mailed initially to 30 or 40 collection managers known to Dessauer and Hafner. Each of these collection managers was asked to identify other collections to be surveyed, and as of May 1983, 155 questionnaires were mailed to scientists throughout the world. We received a total of 113 responses, in which 86 respondents indicated that they maintain collections of viable tissues. Most of the remaining 27 respondents indicated that they use viable tissues in their research, but do not maintain collections.

Following is a summary of the responses of the 86 collection managers identified in our survey. Fifty-five of these collections are located in the United States (representing 26 states and the District of Columbia), and 31 respondents represent 17 foreign countries (see Chapter 12, Directory).

Collection Affiliations

Of the 55 American collections surveyed, 35 (64%) are affiliated with traditional university departments of biology, zoology, or biochemistry. Others are associated with medical or veterinary schools (16%), federal agencies or zoological parks (14%), or private museums (6%). Of the 31 foreign collections identified, 21 (68%) are associated with university departments, five (16%) are affiliated with medical schools, three (10%) are associated with private museums, and two (6%) are maintained by governmental agencies.

Collection Size

Figure 11.1 illustrates the number of U.S. and foreign collections in each of five size categories. The typical U.S. collection ranges in size from several hundred to a few thousand specimens. The typical foreign collection contains approximately

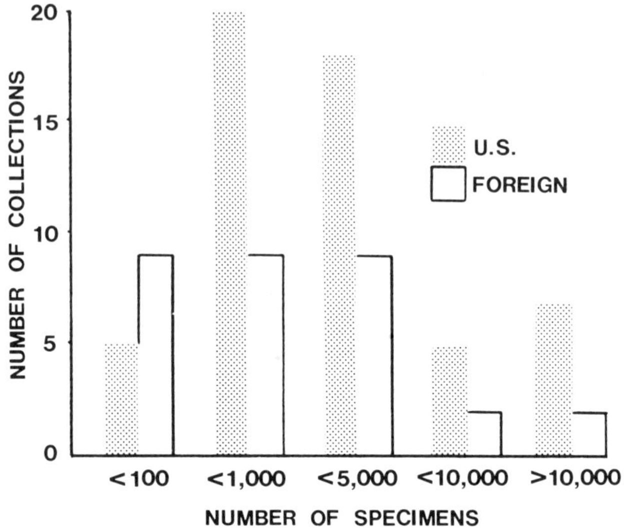

Figure 11.1. Number of specimens (samples) contained in 55 U.S. and 31 foreign collections of frozen tissues.

1,000 specimens. Twelve U.S. collections contain more than 5,000 specimens, and seven of these contain more than 10,000 specimens. The seven largest U.S. collections are located in the states of California, Connecticut, Louisiana, Maryland, New Mexico, and New York. Two foreign collections each contain more than 10,000 specimens; these are located in Argentina and The Netherlands.

Materials Stored in Collections

Ninety-five percent of all collections surveyed (U.S. and foreign) maintain primarily frozen materials. These materials are stored in ultracold deepfreezers in 51% of the U.S. collections and 48% of the foreign collections. The remaining frozen materials are stored in household-type deepfreezers and/or liquid nitrogen.

Nearly all collections surveyed maintain banks of frozen tissues (organs, muscle, etc.) and/or blood. Frozen whole animals, antisera, and isolated proteins are common collection constituents; cell lines and isolated nucleic acids are less common. A few collections contain items such as frozen bile, semen, embryos, saliva, venoms, whole feathers, blood in EDTA, and tissues in EtOH. Chapter 12 contains information on the specimen holdings of each collection surveyed.

When asked how long they maintained collection materials before they are used or discarded, 43% of all managers surveyed responded that they maintain (or hope to maintain) their holdings indefinitely. Thirty-one percent discard their materials after approximately five years, and 26% after one year.

Taxa Represented in Collections

Material contained in the collections surveyed is almost exclusively of animal origin. We were able to identify only two collections, one American and one foreign, that contain significant holdings of viable plant tissues.

Nearly all collections identified in the survey contain specimens of vertebrate animals. Reptile tissues are preserved in 44 of the 86 collections surveyed, mammal material is stored in 42 collections, amphibian tissues are deposited in 32 of the collections, fishes in 28 collections, and birds in 27 collections. Certain collections have large holdings of insects (nine collections), molluscs (six), protozoa (one), and archaebacteria (one).

Geographic Regions Represented

Species from the Nearctic Faunal Region are best represented in the collections surveyed, with 44 of 86 collections indicating that at least 5% of their holdings are from this region. Thirty-three collections have at least 5% of their samples from the Neotropical region, 26 collections have holdings from the Palearctic, 22 from the Oriental region, 20 from the Australian region, and 15 from the Ethiopian region.

Major Sources of Collection Holdings

Sixty-nine percent of U.S. collection managers and 67% of foreign collection managers indicate that they acquire most of their collection materials through their own fieldwork or that of their associates. Twenty-two percent of U.S. collections and 19% of foreign collections obtain most of their holdings from other scientists and/or collections. Commercial sources of viable tissues are important for fewer than 6% of U.S. and 10% of foreign collections. Other sources, including law enforcement agencies, tissue cultures, and zoological parks, were designated as major sources by only three U.S. collections and one foreign collection.

Uses of Collection Materials: Analytical Techniques

Materials preserved in frozen tissue collections are subjected to a wide variety of analytical techniques. By far the most commonly used technique is protein electrophoresis, which is used in 69% of all U.S. collections and 81% of foreign collections. Protein immunological techniques are used in 39% of U.S. collections and 38% of foreign collections. More sophisticated techniques, including protein fingerprinting/sequencing (used in 16% of U.S. collections), nucleic acid analysis (16%), and karyology (4%), are used less commonly.

Uses of Collection Materials: Kinds of Studies

In terms of the kinds of research questions addressed using samples of viable tissues, the survey reveals that these collections are used most heavily for systematic and genetic studies (Figure 11.2). Systematic studies are undertaken in laboratories associated with 73% of both U.S. and foreign collections. Genetic studies are performed in 53% of U.S. collections and 73% of foreign collections. Studies of protein structure are undertaken in 20% of U.S. and 23% of foreign collections. Nucleic acid studies, forensic studies, and a wide variety of other kinds of studies (e.g., hormonal, physiological) are undertaken to lesser degrees in both U.S. and foreign collections.

Collection Curation

As stated above, 95% of the collections surveyed maintain banks of frozen tissues or tissue components. In the event of mechanical or electrical failure, freezers in which the specimens are stored are protected by alarm systems in 79% of the U.S. collections and 42% of foreign collections. Back-up storage systems (usually other

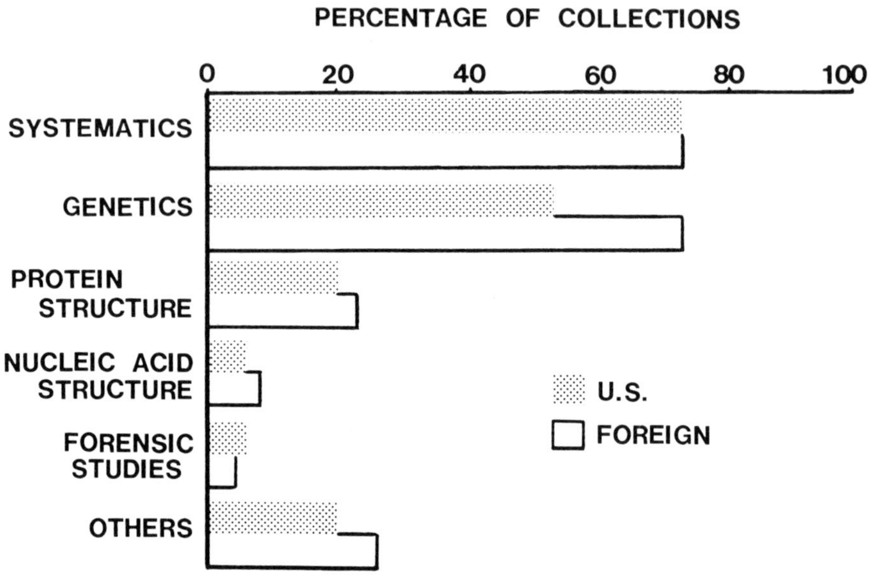

Figure 11.2. Kinds of studies undertaken using tissues and tissue components stored in U.S. and foreign collections of frozen tissues. Materials in many collections are used in a wide variety of studies.

freezers, dry ice, or liquid nitrogen) are available for 90% of the U.S. collections and 85% of the foreign collections.

Only 67% of U.S. collections and 65% of foreign collections maintain a collection catalogue. In other words, of the 86 collections identified, 29 do not maintain a collection catalogue. Those curators who maintain a catalogue usually record: (1) nature of the material; (2) species; (3) collecting locality; and (4) date collected. When asked if they routinely update the catalogue as material is used or given away, 37% of the curators responded "yes," 31% "usually," 16% "sometimes," and 16% "no." Only 69% of the collections in which catalogues are maintained routinely deposit voucher specimens in traditional systematic collections.

The wide variety of materials stored in collections of viable tissues are packaged in many different ways. Plastic cryotubes are commonly used to store tissues, blood, and cell lines. Tissues are also stored in plastic bags, aluminum foil, test tubes, glass vials, plastic wrap, and ice blocks. Blood is also stored in glass vials, hematocrit tubes, eppendorf vials, plastic bags and centrifuge tubes. Whole animals are generally stored in ziplock bags or plastic cryotubes, depending on the size of the specimen. Semen is generally stored in plastic straws or glass vials. Isolated proteins and nucleic acids are usually stored in plastic cryotubes or eppendorf vials.

Nearly all collection managers indicated a willingness to share their material with the scientific community at large. Six of the respondents indicated their willingness to share materials with any interested party. However, the majority of respondents preferred to provide materials only to carefully selected scientists, selection being dependent on the research direction of the potential recipient and the rarity of the specimen (or quantity of material). Four respondents indicated that permits were required to receive their materials. Several collection managers were eager to share their materials, but commented that almost no one knew of the existence of their collection.

Charges for shipping materials from collection to recipient are paid by the donor (collection) 57% of the time, and by the recipient 43% of the time.

Sources of Collection Support

American and foreign collections differ markedly in terms of their major sources of financial support. Individual (basic) research grants to collection managers constitute the major source of support for 47% of all U.S. collections, but only 27% of foreign collections. Conversely, 46% of the foreign collections surveyed receive the bulk of their financial support through university departmental funds. University departmental funds are the major source of support for only 28% of the U.S. collections. The remaining U.S. collections (25%) and foreign collections (27%) receive their support through non-departmental university funds (9% of U.S. collections), federal funds (8%), direct grants to collections (6%), and non-university state funds (2%).

Plans For Long Term Collection Management

The collection managers were asked whether they expect their collections to develop into long term depositories. Twenty-nine (including eight foreign) indicated "yes." Forty-nine (including 20 foreign) responded "no," and eight (including three foreign) were uncertain.

When asked if they favor the development of regional depositories for long term preservation of tissues for use by the scientific community at large, 81% responded "yes," 7% "no," and 12% "not sure" or "doubt feasibility."

When asked if they would deposit their specimens in a regional depository once they completed their work, 91% responded "yes," 1% "no," and 7% "not sure," "too complicated," or "won't be much left over when I finish my work."

CHAPTER 12

DIRECTORY

HOLDINGS IN EXISTING COLLECTIONS OF FROZEN TISSUES

The following is a listing of collections of viable (predominantly frozen) tissues identified in a worldwide survey conducted by H.C. Dessauer and M. S. Hafner between 1981 and 1984. This survey was by no means exhaustive, and we welcome information on collections not listed. We would be especially pleased to receive information on collections of plant tissues; our survey was able to identify only a few such collections, although we assume others exist.

Each listing includes the collection location, person in charge, and specific information concerning collection size, kinds of materials housed in the collection, taxa and geographical regions represented, and major kinds of studies undertaken using the materials in the collection.

Collections are listed alphabetically by state within the United States, then alphabetically by country. Listings are current as of May, 1984.

KEY

SIZE

1 = Less than 100 specimens or samples
2 = Between 100 and 1,000 specimens
3 = Between 1,000 and 5,000 specimens
4 = Between 5,000 and 10,000 specimens
5 = More than 10,000 specimens or samples

MATERIAL

A = Isolated nucleic acids
B = Isolated proteins
C = Cell lines
D = Tissues
E = Blood
F = Antisera
G = Whole animals
H = Other

TAXA

1 = Fishes
2 = Amphibians
3 = Reptiles
4 = Birds
5 = Mammals
6 = Arthropods
7 = Plants
8 = Other

REGIONS

A = Nearctic
B = Palearctic
C = Neotropical
D = Ethiopian
E = Oriental
F = Australian

An asterisk (*) indicates that a material, taxon, or geographic region is proportionately well represented in the collection. A dash (-) indicates that no information is available or that the category is not applicable to the collection.

DIRECTORY

U.S. COLLECTIONS

	SIZE	MATERIALS	TAXA	REGIONS
ARIZONA				
Department of Pharmacology & Toxicology University of Arizona Tucson, AZ 85721 In charge: F. E. Russell (602) 626-4047	3	B,D,E,F,H (venoms) *Pharmacology-Chemistry* (electrophoresis; immunology; protein fingerprinting/sequencing)	1,2,3, 6	A,B,C,E,F
CALIFORNIA				
Museum of Vertebrate Zoology 2593 Life Sciences Bldg. University of California Berkeley, CA 94720 In charge: J. L. Patton (415) 642-3567	5	D*,E*,F *Genetics-Systematics* (electrophoresis; immunology)	2,3,4, 5	A*,B,C,D,F
Department of Zoology University of California Berkeley, CA 94720 In charge: P. Licht	2	B,D,E,F *Protein Structure-Hormone Studies* (immunology; fractionation of proteins)	2,3,4, 5	C
Department of Biology University of California Los Angeles, CA 90024 In charge: D. G. Buth (213) 206-6084	3	D,G *Systematics* (electrophoresis)	1,3	A
Research Department Zoological Society of San Diego San Diego, CA 92112 In charge: O. A. Ryder and K. Benirschke (619) 231-1515	4	C,D,E*, H (sperm, embryos) *Genetics-Cytogenetics-Systematics-Breeding Studies-Pathology-Gene Pool Preservation* (elecrophoresis; karyology; DNA analyses)	3,4,5	A,B,C,D,E, F

COLORADO

	SIZE	MATERIALS	TAXA	REGIONS
Department of Biochemistry Colorado State University Fort Collins, CO 80523 In charge: A. T. Tu (303) 491-6350	1	H (venoms) *Protein Structure* (isolation and characterization of proteins)	3	C,E,F

CONNECTICUT

	SIZE	MATERIALS	TAXA	REGIONS
Biological Sciences Group University of Connecticut Storrs, CT 06268 In charge: A. Brush (203) 486-4560	2	B,H* (feathers) *Protein Structure-Systematics* (electrophoresis; protein fingerprinting/sequencing; supramolecular organization and morphogenesis)	4	
Peabody Museum of Natural History Yale University Box 6666 New Haven, CT 06511 In charge: C. G. Sibley (203) 436-8547	5	A,D,E*,H (egg white) *Systematics-Phylogeny Reconstruction* (hybridization of nucleic acids)	1,2,3 4*,5	A,B,C,D,E, F

FLORIDA

	SIZE	MATERIALS	TAXA	REGIONS
Department of Chemistry Florida State University Tallahassee, FL 32306 In charge: E. Frieden (904) 644-2005	2	E,G *Protein Structure* (electrophoresis; enzyme analysis)	2	
Department of Biochemistry and Molecular Biology University of Florida Medical School Gainesville, FL 32610 In charge: P. J. Laipis (904) 392-6870	2	A*,C,D *Molecular Biology-Evolution* (sequencing, hybridization, and restriction enzyme analysis of nucleic acids)	1,2,3, 4,5*,8	
Dept. of Special Clinical Sciences J. Hillis Miller Health Center P.O. Box J-6 University of Florida Gainesville, FL 32610 In charge: E. R. Jacobson (904) 392-2977	2	D*,E,G *Pathology* (histopathologic evaluation; isolation of infectious agents)	2,3,4, 5	B,C,D,E,F
Department of Biology University of Miami P.O. Box 249118 Coral Gables, FL 33124 In charge: M. M. Miyamoto (305) 284-6206	2	D*,G *Genetics-Systematics* (electrophoresis of proteins)	2,3*	A,C

	SIZE	*MATERIALS*	*TAXA*	*REGIONS*	
Oceanographic Center Nova University 8000 North Ocean Drive Dania, FL 33004 In charge: R. A. Menzies (305) 475-7487	3	D,E,G	1,3*,6, (crustaceans)	A,C	*Genetics-Systematics* (electrophoresis of proteins; immunology)

ILLINOIS

	SIZE	*MATERIALS*	*TAXA*	*REGIONS*	
Field Museum of Natural History Roosevelt Road at Lake Shore Drive Chicago, IL 60605 In charge: H. K. Voris and various investigators (312) 922-9410	2	D,E	2,5*	C*,E	*Systematics-Protein Structure* (electrophoresis)
Biology Department University of Chicago Chicago, IL 60637 In charge: H. B. Shaffer	3	D	2*,3	A,C	*Systematics-Genetics* (electrophoresis; immunology)
Department of Genetics and Development University of Illinois Urbana, IL 61801 In charge: L. Maxson (217) 333-3358	3	E,F	2*,3,5	A,B,C,D, E,F	*Systematics-Genetics-Protein Structure*(immunology)
Department of Genetics and Development University of Illinois Urbana, IL 61801 In charge: G. S. Whitt (217) 333-4645	2	D,F,G*	1		*Genetics-Systematics* (electrophoresis; immunology)

INDIANA

	SIZE	*MATERIALS*	*TAXA*	*REGIONS*	
Department of Microbiology Indiana University School of Medicine 1100 West Michigan Street Indianapolis, IN 46223 In charge: S. Minton (317) 264-7842	2	E*,F	2,3*	A,E,F	*Systematics* (immunology)

LOUISIANA

	SIZE	*MATERIALS*	*TAXA*	*REGIONS*	
Museum of Natural Science Louisiana State University Baton Rouge, LA 70803 In charge: H. C. Dessauer and M. S. Hafner (504) 568-4734 (504) 388-2855	5	B,D*,E,F,H (bile, venom)	1,2,3*, 4*,5,8	A*,B,C*, D,E,F	*Genetics-Systematics-Protein Structure Forensic Problems* (electrophoresis; immunology; protein and DNA fingerprinting)

	SIZE	*MATERIALS*	*TAXA*	*REGIONS*
Department of Biochemistry Louisiana State University Medical Center 1901 Perdido Street New Orleans, LA 70112 In charge: R. E. Reeves (504) 568-4734	2	G	8 (amoebae)	A,C
	Enzymology (enzyme analyses)			

MARYLAND

	SIZE	*MATERIALS*	*TAXA*	*REGIONS*
Cell Culture Department American Type Culture Collection 12301 Parklawn Drive Rockville, MD 20852 In charge: R. J. Hay (301) 881-2600	5	C*,D,E,F	1,2,3,4, 5*,6,8	
	International Repository for Cell Lines (lines characterized using cytogenetics, isoenzymology, and other biochemical tests; immunology)			
Section of Genetics National Cancer Institute Building 560, Room 11-85 Frederick, MD 21701 In charge: S. J. O'Brien	3	A,B,C, D,E	5	
	Genetics-Systematics (electrophoresis; immunology; hybridization of nucleic acids)			
U.S. Fish and Wildlife Service Patuxent Wildlife Research Center Laurel, MD 20811 In charge: G. F. Gee (301) 498-0419	2	H (avian semen)	4	
	Preservation of Endangered Species (frozen semen gene pool)			
Department of Zoology University of Maryland College Park, MD 20742 In charge: R. Highton (301) 454-6742	5	E,G* (homogenates)	2*,3	A
	Systematics (electrophoresis)			

MASSACHUSETTS

	SIZE	*MATERIALS*	*TAXA*	*REGIONS*
Department of Entomology University of Massachusetts Amherst, MA 01003 In charge: S. R. Diehl (413) 545-2284	3	G	6 (Diptera: Tephritidae)	A
	Systematics-Genetics-Applied Problems in Agriculture(electrophoresis; immunology; m-DNA restriction enzyme analysis)			

MICHIGAN

	SIZE	*MATERIALS*	*TAXA*	*REGIONS*
Division of Biological Sciences University of Michigan Ann Arbor, MI 48109 In charge: W. M. Brown (313) 763-0497	3	A,C,D	3*,5,8	A*,C,D,E
	Genetics-Systematics-Nucleic Acid Structure(sequencing, hybridization, and restriction endonuclease analysis of DNA)			

	SIZE	*MATERIALS*	*TAXA*	*REGIONS*
Department of Anatomy Wayne State University School of Medicine Detroit, MI 48201 In charge: M. Goodman (313) 577-1004	4	A,B,D* E*,F *Systematics-Genetics* (electrophoresis; immunology; protein fingerprinting/sequencing; sequencing of nucleic acids)	3,4,5*	A,C,D*,E*

MINNESOTA

	SIZE	*MATERIALS*	*TAXA*	*REGIONS*
James Ford Bell Museum of Natural History University of Minnesota Minneapolis, MN 55455 In charge: K. W. Corbin (612) 373-5643	3	B,D*,E, G* *Systematics-Genetics* (electrophoresis; protein fingerprinting/sequencing)	4,6 (scale insects)	A,C

MONTANA

	SIZE	*MATERIALS*	*TAXA*	*REGIONS*
Department of Zoology University of Montana Missoula, MT 59812 In charge: K. Knudsen and F. Allendorf (406) 243-5503	3	D,G *Genetics-Systematics* (electrophoresis)	1	A

NEW JERSEY

	SIZE	*MATERIALS*	*TAXA*	*REGIONS*
Bureau of Biological Research Nelson Biological Laboratory Rutgers University, P.O. Box 1059 Piscataway, NJ 08854 In charge: R. C. Vrijenhoek (201) 932-2804	3	G *Genetics-Systematics* (electrophoresis)	1*,8	A*,D
Department of Biological Sciences Nelson Biological Laboratory Rutgers University Piscataway, NJ 08854 In charge: D. E. Fairbrothers (201) 932-2843	2	H (seeds, pollen) *Systematics* (electrophoresis, immunochemistry, serology)	7	A

NEW MEXICO

	SIZE	*MATERIALS*	*TAXA*	*REGIONS*
Museum of Southwestern Biology University of New Mexico Albuquerque, NM 87131 In charge: T. L. Yates (505) 277-3838	5	C,D,E *Systematics-Genetics* (electrophoresis)	1,2,3 4,5*	A*,B,C,E

NEW YORK

	SIZE	MATERIALS	TAXA	REGIONS
American Museum of Natural History Central Park West at 79th Street New York, NY 10024 In charge: G. Barrowclough (212) 873-1300, ext.423	3	D,E	4	A,C
	Systematics-Genetics (electrophoresis)			
Division of Biological Sciences Section of Ecology and Systematics Cornell University Ithaca, NY 14850 In charge: P. F. Brussard (607) 256-6582	2	G,H (insect homogenates)	6	A*,B,C
	Genetics-Systematics (electrophoresis)			
Department of Biological Sciences Fordham University Bronx, NY 10458 In charge: R. C. Dowler	2	D,E	5	A
	Systematics-Genetics (electrophoresis of proteins)			
Department of Biology The King's College Briarcliff Manor, NY 10510 In charge: W. Frair (914) 941-7200	3	A,B,E*, F,H (fat)	1,2,3* 4,5 (turtles)	A*,B,C,D, E,F
	Systematics (electrophoresis; immunology)			
The Mary Imogene Bassett Hospital Cooperstown, NY 13326 In charge: T. Peters, Jr. (607) 547-3670	1	E	1,2,3 4,5	C*,E
	Protein Structure (electrophoresis; protein fingerprinting/sequencing; protein ligand binding)			
Roswell Park Memorial Institute 666 Elm Street Buffalo, NY 14263 In charge: E. Cohen (716) 845-5778	3	C*,E,F	5 (human)	
	Comparative and Clinical Immunology; Histocompatibility; Blood banking (immunology)			
Department of Biology University of Rochester Rochester, NY 14627 In charge: H. Ochman (716) 275-4868	5	B*,G (snails, slugs)	8	A,B (Europe)
	Genetics (electrophoresis)			

NORTH CAROLINA

	SIZE	MATERIALS	TAXA	REGIONS
Biology Department Belmont Abbey College Belmont, NC 28012 In charge: M. J. McLeod (704) 825-3711	1	G	8 (Mollusca)	A
	Genetics-Systematics (electrophoresis)			

OHIO

	SIZE	*MATERIALS*	*TAXA*	*REGIONS*
Cincinnati Zoological Gardens 3400 Vine Street Cincinnati, OH 45220 In charge: B. L. Dresser (513) 281-4701	2	H (sperm, embryos)	2,4,5*	A*,B,C,D, E,F
	Artificial Insemination; Embryo Transfer			

PENNSYLVANIA

	SIZE	*MATERIALS*	*TAXA*	*REGIONS*
Philadelphia Academy of Natural Sciences 19th and The Parkway Philadelphia, PA 19103 In charge: T. Post and T. Uzzell (215) 299-1190	3	A,C,E,F G*	2*,3	A,B*,E
	Systematics (immunology; electrophoresis; restriction analysis of m-DNA)			

SOUTH CAROLINA

	SIZE	*MATERIALS*	*TAXA*	*REGIONS*
National Marine Fisheries Service U.S. Department of Commerce Charleston Laboratory P.O. Box 12607 Charleston, SC 29412 In charge: S. A. Braddon (803) 724-4762	1	B,D*,G*	1,3	A (SE U.S.)
	Protein Structure-Forensic Problems (electrophoresis; contaminant analysis)			
Savannah River Ecology Laboratory Institute of Ecology P. O. Drawer E Aiken, SC 29801 In charge: M. H. Smith (803) 725-2472	4	D	1,5	A
	Ecology-Genetics-Systematics (protein electrophoresis)			

TENNESSEE

	SIZE	*MATERIALS*	*TAXA*	*REGIONS*
Department of Zoology University of Tennessee Knoxville, TN 37916 In charge: G. F. McCracken and K. Gustin (615) 974-3699; 974-6194	4	D,E*,G	1,3,5* 6,8 (gastropods)	A,C,D
	Genetics-Systematics-Forensic Problems (electrophoresis)			

TEXAS

	SIZE	*MATERIALS*	*TAXA*	*REGIONS*
Department of Biology Texas A&M University College Station, TX 77843 In charge: I. F. Greenbaum (409) 845-7791	3	C,D*	1,3,5*	A,C*
	Genetics-Systematics-Fisheries Management (electrophoresis; karyology)			

	SIZE	*MATERIALS*	*TAXA*	*REGIONS*
Department of Biological Sciences Texas Tech University Lubbock, TX 79409 In charge: R. J. Baker (806) 742-2485	4	C,D*	3,5*	A,C*,D,F
	Genetics-Systematics (karyology; elecrophoresis; immunology)			
Department of Biological Sciences Texas Tech University Lubbock, TX 79409 In charge: F. L. Rose (806) 742-2726	2	B,D*,E,F	3 (turtles)	A
	Systematics (electrophoresis; immunology)			
Department of Biology SR2 University of Houston Houston, TX 77004 In charge: D. L. Jameson (713) 749-1479	2	A,D,G	1,2*,3	A
	Genetics-Systematics (electrophoresis; sequencing of nucleic acids)			
Department of Zoology University of Texas Austin, TX 78712 In charge: A. F. Riggs (512) 471-1585	2	B,E	1,2	A,C
	Protein and Nucleic Acid Structure (protein fingerprinting/sequencing; sequencing of nucleic acids)			

VERMONT

	SIZE	*MATERIALS*	*TAXA*	*REGIONS*
Department of Zoology University of Vermont Burlington, VT 05405 In charge: C. W. Kilpatrick (802) 656-2922	3	B,C,D*,E F,H (saliva)	3,5*	A*,C
	Systematics-Genetics (electrophoresis; immunology; karyology)			

WASHINGTON

	SIZE	*MATERIALS*	*TAXA*	*REGIONS*
National Marine Fisheries Service 2725 Montlake Blvd. East (Pacific Ocean) Seattle, WA 98112 In charge: Fred Utter (206) 442-2737; 842-5832	2	D,E,F,G	1	B
	Genetics-Systematics (electrophoresis; immunology)			

WASHINGTON, D.C.

	SIZE	*MATERIALS*	*TAXA*	*REGIONS*
Department of Pathology National Zoological Park Smithsonian Institution Washington, D.C. 20008 In charge: R. A. Freeman and D. Fischer	3	D*,G,H* (sera, semen)	2,3,4 5*	B,C*,D, E,F
	Forensic Pathology-Virology The Department of Zoological Research uses tissues for studies in *Systematics-Anatomy-Endocrinology*			

	SIZE	*MATERIALS*	*TAXA*	*REGIONS*

WEST VIRGINIA

Department of Biological Sciences Marshall University Huntington, WV 25701 In charge: M. E. Seidel (304) 696-2427	2	D	3 (turtles)	A,C*
	Systematics (electrophoresis)			

WISCONSIN

Department of Physiological Chemistry University of Wisconsin Medical School Madison, WI 53706 In charge: W. M. Fitch (608) 262-1475	1	B,D*,E,G	2,3,5, 8	
	Systematics-Protein Structure (protein fingerprinting/sequencing)			
Department of Life Sciences University of Wisconsin-Parkside Kenosha, WI 53140 In charge: J. S. Balsano (414) 553-2475	2	B,E,G	1	C
	Genetics (electrophoresis; immunology)			

FOREIGN COLLECTIONS

ARGENTINA

	SIZE	*MATERIALS*	*TAXA*	*REGIONS*
Instituto de Biologia Animal Universidad Nacional de Cuyo Mendoza, Argentina In charge: L. R. Castro	5	D*,E,F	1,2*,3	C
	Genetics-Systematics-Protein Structure (electrophoresis; immunology; karyology)			

AUSTRALIA

	SIZE	*MATERIALS*	*TAXA*	*REGIONS*
Department of Zoology Macquarie University North Ryde New South Wales 2113 Australia In charge: S. Donnellan	2	C,D,E	2,3*,4	F
	Genetics-Systematics (electrophoresis; hybridization of nucleic acids)			
CSIRO Marine Laboratories P.O. Box 120, Cleveland Queensland 4163, Australia In charge: J. B. Shaklee	3	D*,F,G	1	F
	Genetics-Systematics (electrophoresis of proteins)			
South Australian Museum North Terrace, Adelaide South Australia 5000 Australia In charge: P. R. Baverstock	3	C,D*,E,F	2,3,4, 5*	F
	Genetics-Systematics (electrophoresis; immunology)			
South Australian Museum North Terrace, Adelaide South Australia 5000 Australia In charge: T. D. Schwaner	3	D,E,F,H, (DNA)	3,5,8	F
	Genetics-Systematics (electrophoresis; immunology)			
Department of Zoology University of New South Wales P.O. Box 1, Kensington New South Wales 2033, Australia In charge: R. H. Crozier	3	D,G*	4,6*,8 (insects)	F
	Genetics-Systematics-Nucleic Acid Structure (electrophoresis; m-DNA analysis)			

AUSTRIA

Institution	*SIZE*	*MATERIALS*	*TAXA*	*REGIONS*
Institute of Medical Chemistry Veterinary Medical University Vienna, Austria 1030 In charge: H. Czikeli	1	D*,E,F,G	4	B
	Genetics-Systematics-Protein and Nucleic Acid Structure-Forensic Problems (electrophoresis; isoelectric focusing)			

CANADA

Institution	*SIZE*	*MATERIALS*	*TAXA*	*REGIONS*
University of Toronto and Royal Ontario Museum 100 Queen's Park Toronto Ontario, Canada M5S 2C6 In charge: K. E. Chua and various investigators	2	D*,E,G	1,4*	A,B,C,F
	Genetics-Systematics (electrophoresis)			

CHINA

Institution	*SIZE*	*MATERIALS*	*TAXA*	*REGIONS*
Institute of Zoology Academia Sinica Taipei, Taiwan 115 Republic of China In charge: K. Y. Jan	1	C,E	5	E
	Cytogenetics-Carcinogenesis			
Department of Biomorphics National Defense Medical Center P.O. Box 8244 Taipei, Taiwan 107 Republic of China In charge: C. Y. Hsu	1	D,F	2	E
	Developmental Endocrinology (immunology; enzyme analysis)			
Department of Biomorphics National Defense Medical Center P.O. Box 8244 Taipei, Taiwan 107 Republic of China In charge: S. Mao	1	E*,F	3	E
	Systematics (electrophoresis; immunology; protein fingerprinting/sequencing)			

SIZE *MATERIALS* *TAXA* *REGIONS*

ENGLAND

Department of Biology
University of Salford
Salford M5 4WT
United Kingdom
In charge: R. Lawson
and technician

SIZE	MATERIALS	TAXA	REGIONS
2	D,E*,F	1	

Genetics-Systematics-Protein Structure
Muscle Biochemistry (electrophoresis; immunology; histology)

Department of Human Sciences
University of Technology
Loughborough, Leicestershire
United Kingdom
In charge: R. D. Ward

SIZE	MATERIALS	TAXA	REGIONS
1	G	1	B

Genetics-Systematics
(electrophoresis)

FINLAND

Finnish Game and Fisheries
Research Institute
Riista-ja Kalatalouden
Tutkimuslaitos
Ahvenjarven Riistantutkimusasema
82950 Kuikkalampi
Finland
In charge: T. Nygren

SIZE	MATERIALS	TAXA	REGIONS
4	D*,E	4,5*	B

Genetics-Forensic Problems
(electrophoresis of proteins)

FRANCE

Department of Zoogeographie
Universite Paul Valery
Montpellier
France
In charge: F. Blanc

SIZE	MATERIALS	TAXA	REGIONS
2	D*,E,F	3*,4,8 (molluscs)	B*,D,F

Systematics-Genetics
(electrophoresis; immunology)

IRELAND

Zoology Department
Queen's University
Belfast BT7 1NN
N. Ireland
In charge: A. Ferguson

SIZE	MATERIALS	TAXA	REGIONS
3	D*,G	1	A,B*

Systematics-Genetics
(electrophoresis)

	SIZE	*MATERIALS*	*TAXA*	*REGIONS*

ISRAEL

Institution	SIZE	MATERIALS	TAXA	REGIONS
Laboratory of Population Biology Institute of Evolution Haifa University Haifa Israel In charge: E. Nevo	3	D,G, (homogenates)	2,3,5,7,8 (shrimp, molluscs, wild wheat, barley)	B
	Genetics-Systematics-Ecology (electrophoresis)			

JAPAN

Institution	SIZE	MATERIALS	TAXA	REGIONS
Department of Chemistry Tohoku University Aobayama, Sendai Japan In charge: N. Tamiya	2	B,D,H, (venoms)	3	E,F
	Protein Structure and Function (protein fingerprinting/sequencing)			

MALAYSIA

Institution	SIZE	MATERIALS	TAXA	REGIONS
Department of Biology University of Agriculture Serdang, Selangor Malaysia In charge: G. Y. Yuen and lab assistant	2	D,E	1,5*	E
	Genetics (electrophoresis)			
Department of Genetics and Cellular Biology University of Malaya Kuala Lumpur Malaysia In charge: H. S. Yong and technician	2	D,E,F,G	1,2,3, 5*,6* (insects)	E
	Genetics-Systematics (electrophoresis; immunology)			

NETHERLANDS

Institution	SIZE	MATERIALS	TAXA	REGIONS
Institute of Human Genetics Faculty of Medicine Free University P.O. Box 7161 1007 MC Amsterdam The Netherlands In charge: R. R. Frants	3	E,H (urine, saliva)	5 (human)	B
	Genetics (electrophoresis)			

	SIZE	MATERIALS	TAXA	REGIONS
Department of Human Genetics University of Leiden Wassenaarseweg, 72 2333 AL Leiden The Netherlands In charge: P. M. Khan and collaborators	5	C,D,E*	5	
	Genetics-Systematics-Protein Structure and Function (electrophoresis; immunology; enzymology)			
Department of Biochemistry University of Nijmegen The Netherlands In charge: W. W. deJong	1	A,B,D (eyes, lenses)	3,4,5	
	Protein Structure-Nucleic Acid Structure (Protein and nucleic acid sequencing)			

NORWAY

	SIZE	MATERIALS	TAXA	REGIONS
Zoological Institute Agricultural University of Norway 1432 As-NLH Norway In charge: K. Røed	2	D,E	5	A,B*
	Genetics-Systematics (electrophoresis)			

POLAND

	SIZE	MATERIALS	TAXA	REGIONS
Department of Comparative Anatomy Jagiellonian University Krakow, Poland In charge: J. M. Szymura and J. Rafinski	2	D*,E,G	2	B
	Genetics-Systematics (electrophoresis)			

SWITZERLAND

	SIZE	MATERIALS	TAXA	REGIONS
Institute of Zoology University of Berne Baltzerstr. 3 CH-3012, Berne Switzerland In charge: H. J. Geiger and assistant	4	G	6	A,B*
	Systematics (electrophoresis)			
Institut de Zoologie Universite de Lausanne CH-1015 Lausanne-Dorigny Switzerland In charge: F. Catzeflis and T. Maddalena	3	D*,E,F	2,5*,6, 8 (snails)	A,B*
	Systematics (electrophoresis; immunology)			

	SIZE	*MATERIALS*	*TAXA*	*REGIONS*
Institut fur Pathologie Universitatsspital Sternwartstrasse 2 8091 Zurich Switzerland In charge: H. Hengartner and R. M. Zinkernagel	1	C	5 (lab mice)	
	Genetics-Protein and Nucleic Acid Structure (immunology)			

WEST GERMANY

	SIZE	*MATERIALS*	*TAXA*	*REGIONS*
Fachbereich Biologie der Philipps-University Lahnberge 3550 Marburg West Germany In charge: U. Joger	3	B,D,E*,F, G	3,8 (Crustacea/Mollusca)	B*,C,D
	Systematics (electrophoresis; immunology)			
Institut fur Physiologie Universitat Regensburg 8400 Regensburg West Germany In charge: C. Bauer	1	B	3,5	
	Respiratory Function of Hemoglobin			
Max-Planck-Institut fur Biochemie 8033 Martinsried bei Munchen West Germany In charge: T. Kleinschmidt and G. Braunitzer	1	B,D	2,3,4, 5	
	Protein Structure (protein sequencing; hemoglobin studies)			

SECTION V

PROPOSED PLAN FOR THE ORGANIZATION AND SUPPORT OF U.S. FROZEN TISSUE COLLECTIONS

The Workshop Panel

Section I of this report has documented at length the value of frozen tissue collections to science and society. The areas of basic and applied research that benefit directly from frozen tissue collection resources are as diverse as evolutionary biology, veterinary and medical science, and forensic and environmental research. Tissues from more than one-quarter million organisms are currently stored in frozen collections throughout the United States, and collection holdings are increasing at a rapid rate. These holdings, viewed collectively, have now attained the status of a National Resource, albeit one that is diffuse and largely unmanaged. To assure the continued existence of this fragile resource, coordinate its growth and support, and promote its effective use, the Workshop Panel recommends that a carefully conceived national management plan for frozen tissue collections be implemented. A Council of the Association of Systematics Collections should be established to refine and promote this plan. This Council should work in close harmony with program officers of the National Science Foundation and National Institutes of Health concerned with the development and preservation of national biological resources. The Workshop Panel recommends that research funding agencies give careful consideration to the following proposed National Plan.

Designation of *Frozen Tissue Depositories*

Many existing frozen tissue collections in the U.S. consist of small accumulations of valuable tissues collected for a single research project. Often, because of limited space or funds, precious material is discarded upon completion of the project or termination of grant support. Many individual scientists who lack the resources and/or inclination to manage long term tissue banks, nevertheless recognize the value of their material and are willing to deposit it where it will be preserved for future use (see Chapter 11).

At present, no single collection has the facilities, funding, or manpower to act as a central depository. For this reason, the Workshop Panel recommends that a small number of existing frozen tissue collections be designated as *Frozen Tissue Depositories*, whose principal mission would entail long term preservation of tissue resources for the research community at large. Because these depositories would function in the national interest, they should be assured of stable and continuing federal support.

Location of *Frozen Tissue Depositories*

The Workshop Panel recommends that approximately five *Frozen Tissue Depositories* be established initially. These should be selected from among those institutions that already have large collections, considerable capital equipment, and a curatorial staff that has a demonstrated commitment to long term tissue preservation. The Wake

Committee (Wake et al., 1975), which recommended the establishment of centralized tissue depositories nearly a decade ago, suggested that the most appropriate sites for such depositories would be university natural history museums. Because of the presence of experts in a wide variety of physical, chemical, and biological disciplines, the university environment would promote broad use of collections and assist curators in keeping pace with technological advances in cryopreservation techniques.

Many of the 55 U.S. frozen tissue collections identified in the survey (see Chapter 12) qualify as possible sites for designated depositories. Of the larger collections, those located at the following universities would seem to be prime candidates (in alphabetical order by state): University of California at Berkeley, University of Illinois at Urbana, Louisiana State University, Wayne State University, University of Minnesota, University of New Mexico, and Texas Tech University. In addition, the collection maintained by the Zoological Society of San Diego would seem to be an appropriate candidate because of its size, quality, and the nature of its research commitment.

For a collection to be recognized as a *Frozen Tissue Depository*, the curator should be required to submit a proposal to an appropriate federal granting agency describing the facility, its existing staff and collection holdings, and providing tangible evidence of the institution's commitment to long term financial support for the depository (see beyond). The curator must provide convincing evidence that, if the collection at his institution is selected as a *Frozen Tissue Depository* supported by federal funds, he will accept, preserve, and distribute tissues for the scientific community at large. The proposal should document a clear commitment to service, with funds requested solely for collection curation.

Personnel of Depositories

Frozen Tissue Depositories should be managed by research scientists with special interest and competence in tissue cryopreservation techniques and extensive training and experience in macromolecular chemistry. The collection manager should hold professional status in the institution housing the facility. Besides the responsibility for maintaining high curatorial standards, such as careful documentation of source of specimens, the scientist should be using the material in the collection in his own research.

To aid in the multiple tasks of processing tissues, cataloging specimens, and maintaining vital equipment, at least one full time technician should be employed. Other technical help may be necessary to maintain equipment and perform routine analytical work on the collection.

To take full advantage of local expertise, a campus committee should be formed to advise the curator regarding development and management of the collection. This committee should include scientists selected from such university departments as biochemistry, microbiology, physics, and engineering.

Financial Support For *Frozen Tissue Depositories**

Because frozen tissue resources are useful in so many different areas of research, it is not surprising that no single research funding agency has seen fit to assume sole

*Contributed in large part by Gregory S. Whitt

responsibility for their long term support. Until now, frozen tissue collection managers have made no coordinated effort to relate the value of their collections to the missions of federal, state, or private granting agencies; as a result, nearly all tissue collections in the U.S. have received their support only indirectly, with minimal curatorial costs often buried within the budgets of research grants to individual investigators (see Chapter 11). As collections have grown in size and curatorial costs have increased, the funding base has remained low and undependable; as a result many valuable collections have been lost and others are in jeopardy.

Today, America's frozen tissue collections have come of age; they are, without doubt, a valuable biological resource deserving of stable, broad-based, and continuing support. The institutions housing frozen tissue collections, and many federal, state, and private agencies have reaped benefits from these resources, often far exceeding the level of financial investment. The Workshop Panel has concluded that all of the following institutions and agencies share the responsibility of supporting the proposed *Frozen Tissue Depositories*.

Responsibilities of the Home Institution.–The many institutions across the nation that house frozen tissue collections have, to date, borne the major financial burden for collection support. The Workshop Panel believes it appropriate that the home institutions continue to provide the physical facilities (space, power, water) and service functions (secretarial and janitorial) for the collections. These are not trivial support items. Perhaps most importantly, the home institution should be expected to demonstrate a long term commitment to its frozen tissue collection by funding the salary of a full time professional Curator of Frozen Tissues. To our knowledge, only one collection in the nation has such a university-funded position. "Hard money" support for this curatorial position with university funds is critical for the long term development of the depository.

Responsibilities of Federal Agencies Supporting Biological Research Resources.– Indirectly, through its support of fieldwork, the Systematic Biology Program of the National Science Foundation can accept most of the credit for the present size and diverse nature of the U.S. frozen tissue resource (see Chapter 11). The Systematic Biology Program should continue to support basic research that contributes specimens to frozen tissue collections; however, the Workshop Panel has concluded that the major responsibility for support of *Frozen Tissue Depositories* belongs more properly with the NSF Biological Research Resources Program, which has furnished support in the past for at least two frozen tissue collections. In addition, the responsibility for support of these depositories should be shared by comparable programs of the National Institutes of Health, whose grantees have utilized these collections in the past and will use them with increasing frequency in the future (see Chapters 2 and 3).

Biological resource support programs of both the National Science Foundation and the National Institutes of Health should share costs for the purchase of capital equipment (e.g., ultracold freezers, alarm systems), curatorial supplies (e.g., cryotubes, liquid nitrogen), equipment repair, and shipment of tissues to and from the depository. These agencies should provide the salary for at least one support person in each *Frozen Tissue Depository*.

Because the institutions most likely to be designated as tissue depositories already have much of the necessary capital equipment, equipment costs should be far less than one might envision. Primary capital needs would include occasional money to purchase additional ultracold freezers to increase tissue storage capacity and instrumentation needed to process tissues.

Responsibilities of Other Funding Agencies.–Because of the broad value of the frozen tissue resource, many funding agencies, in addition to NSF and NIH, should participate in the development of *Frozen Tissue Depositories*. This participation would be mostly indirect, through support of research concerned with cryopreservation techniques and funding of field projects that contribute tissues to collections. Funding agencies should encourage investigators to utilize the network of depositories whenever possible. When feasible, grantees should use tissues already preserved in a depository, thus eliminating some unnecessary fieldwork. Funding agencies should also encourage grantees to donate specimens to a *Frozen Tissue Depository* on completion of research projects. These steps would reduce waste of animal and plant resources and allow for maximal use of each specimen collected.

Although the missions of funding agencies vary considerably, the aims of many would be advanced through their support of *Frozen Tissue Depositories*. For example, the National Institutes of Health should appreciate the importance of cryopreserved tissues for medical research. A persuasive case could be made for preserving both normal and diseased tissues from a wide variety of species for subsequent use in retrospective studies (see Chapter 3). Many research programs sponsored by the U.S. Department of Agriculture would be aided by the availability of cryopreserved tissues from plant and animal species of economic importance.

Various regulatory agencies, including the Environmental Protection Agency, the Office of Marine Pollution of NOAA, and the National Marine Fisheries Services, would undoubtedly find *Frozen Tissue Depositories* valuable for preserving comprehensive samplings of tissues collected from species inhabiting polluted environments. Retrospective genetic and physiological studies using these samples would yield important information on the long term effects of environmental perturbation.

The Dingell-Johnson and Pittman-Robertson funds distributed by the U.S. Fish and Wildlife Service and earmarked for state wildlife projects should support research that involves the collection of tissues from game species for use in management programs and forensic studies. The Office of Naval Research should recognize the need for developing collections of frozen tissues from sharks, cetaceans, pinnipeds, and other marine organisms whose biology may have direct or indirect human health implications. Both the Office of Naval Research and the Air Force Office of Scientific Research should be eager to support studies involving the collection of tissues and venoms from poisonous organisms for use in the development of antivenins.

Conclusions.—In the final analysis, the responsibility for stable, continuing support for *Frozen Tissue Depositories* should be shared by the home institution and the biological resource programs of the National Science Foundation and the National Institutes of Health. These, and other agencies funding biological research (Escherich and McManus, 1983), should support the development and expansion of *Frozen Tissue Depositories* to the extent that these collections assist them in fulfilling their missions.

SECTION VI

LITERATURE CITED

Academy of Sciences, USSR. 1980. Genome Conservation. Moscow, USSR.

Anderson, J. O., J. Nath and E. J. Harner. 1978. Effect of freeze-preservation on some pollen enzymes. Cryobiology, 15: 469–477.

Aquadro, C. F., and B. D. Greenberg. 1983. Human mitochondrial DNA variation and evolution: Analysis of nucleotide sequences from seven individuals. Genetics, 103: 287–312.

Baker, C. M. A. 1966. Species, tissue, and individual specificity of low ionic strength extracts of avian muscle and other organs revealed by starch gel electrophoresis. Canadian J. Biochem., 44: 853–859.

Bendz, G., and J. Santesson (eds.). 1974. Chemistry in botanical classification. Proc. 25th Nobel Symp., Academic Press, New York.

Benirschke, K. 1982. Cell and tissue banks. Proc. Amer. Assoc. Zool. Parks and Aquaria, 1982, pp. 154–158.

Benirschke, K., and M. Bogart. 1978. Pathology of Douc langurs *Pygathrix nemaeus* at the San Diego Zoo. XXth Int'l Symp. Erkr. Zootiere, Akademie Verlag, Berlin, pp. 257–261.

Benirschke, K., and A. T. Kumamoto. 1983. Paternity diagnosis in pygmy chimpanzees. Int'l Zoo Yearbook. In press.

Bishop, J. A., and L. M. Cook (eds.). 1981. Genetic Consequences of Man Made Change. Academic Press, New York. 409 pp.

Blumberg, B. S., and I. Warren. 1961. The effect of sialidase on transferrins and other serum proteins. Biochim. Biophy. Acta, 50: 90–101.

Boyden, A. 1948 through 1978. Serological Museum Bulletin. Vols. 1 through 51, Rutgers Univ., New Brunswick, New Jersey.

Brazaitis, P., and M. Watanabe. 1982. The doppler, a new tool for reptile and amphibian hematological studies. J. Herpetology, 16: 1–6.

Britten, R. J., and D. E. Kohne. 1968. Repeated sequences in DNA. Science, 161: 529–540.

Brown, W. M. 1980. Polymorphism in mitochondrial DNA in humans as revealed by restriction endonuclease analysis. Proc. Nat. Acad. Sci., USA, 77: 3605–3609.

Brown, W. M., M. George, and A. C. Wilson. 1979. Rapid evolution of animal mitochondrial DNA. Proc. Nat. Acad. Sci., USA, 76: 1967–1971.

Bryson, V., and H. J. Vogel (eds.). 1965. Evolving Genes and Proteins. Academic Press, New York. 629 pp.

Bunn, H. F., and P. J. Higgins. 1981. Reaction of monosaccharides with proteins: Possible evolutionary significance. Science, 213: 222–224.

Cambell, C. A. 1979. Genetic divergence between populations of *Thais lamellosa* (Gmelin). pp. 157–170 *In*: B. Battaglia and J. A. Beardmore (eds.), Marine Organisms, Genetics, Ecology, and Evolution. Plenum Press, New York.

Champion, A. B., E. M. Prager, D. Wachter and A. C. Wilson. 1974. Microcomplement Fixation. pp. 397–416 *In*: C. A. Wright (ed.), Biochemical and Immunological Taxonomy. Academic Press, New York.

Conway, W. G. 1980. An overview of captive propagation. pp. 199–208 *In*: M. F. Soule, and B. A. Wilcox (eds.), Conservation Biology. Sinauer Publ., Sunderland, Maryland.

Davies, G. (ed.). 1975. Forensic Science. ACS Symp. 13, Amer. Chem. Soc., Washington, D.C. 204 pp.

Dawson, D. M., H. M. Eppenberger, and N. O. Kaplan. 1967. The comparative enzymology of creatine kinases. II. Physical and chemical properties. J. Biol. Chem., 25: 210–217.

Dayhoff, M. O., L. T. Hunt, W. C. Barker, B. C. Orcutt, L. S. Yeh, H. R. Chen, D. G. George, M. C. Blomquist, and G. C. Johnson. 1983. Atlas of Protein Sequence and Structure. Version 6. Nat. Med. Res. Foundation, Georgetown Univ. Med. Ctr., Washington, D.C.

DeMay, D. J., and R. A. Menzies. 1982. Evidence for cytochrome P-450 mixed function oxidases in marine algae. Fed. Proc., 41: 1298.

Dessauer, H. C. 1970. Blood chemistry of reptiles. pp. 1–72 *In*: C. Gan and T. S. Parsons (eds.), Biology of the Reptilia, Vol. 3. Academic Press, New York.

Dessauer, H. C., M. J. Braun, and S. Neville. 1983. A simple hand centrifuge for field use. Isozyme Bulletin, 16: 91.

D'Eustachio, P. 1984. Gene mapping and oncogenes. Amer. Scientist, 72: 32–40.

Doellgast, G. J., and K. Benirschke. 1979. Placental alkaline phosphatase in Hominidae. Nature, 280: 601–602.

Escherich, P. C., and R. E. McManus (eds.). 1983. Sources of Federal Funding for Biological Research: A Biologist's Guide to Successful Fund Raising Procedures. Assoc. of Systematics Collections, Lawrence, Kansas. 84 pp.

Fairbrothers, D. E. 1969. Plant serotaxonomy (serosystematic literature), 1951–68. Serological Museum Bulletin, Rutgers Univ., 41: 1–10.

Fennema, O., and J. C. Sung. 1980. Lipoxygenase-catalyzed oxidation of linolenic acid at subfreezing temperatures. Cryobiology, 17: 500–507.

Fitch, W. M. 1982. The challenges to Darwinism since the last centennial and the impact of molecular studies. Evolution, 36: 1133–1143.

Gadi, I. K., and O. A. Ryder. 1983. Distribution of silver-stained nucleolus organizing regions in the chromosomes of the Equidae. Genetica, 62: 109–116.

Gardner, M., P. Marx, D. Maul, K. Osborn, L. Lowensteine, N. Lerche, R. Hendrickson, B. Munn, B. Belncken, M. Bryant and J. Sever. 1984. Simian acquired immune deficiency syndrome: An overview. *In*: M. S. Gotlieb, and J. E. Groopman (eds.), UCLA Symposia on Molecular and Cellular Biology. New Series, Vol. 16. Alam R. Liss, Inc., New York, NY. In Press.

Gee, G. F. 1983. Avian artificial insemination and semen preservation. pp. 375–398 *In*: Proc. Jean Delacour/IFCB Symposium on Breeding Birds in Captivity. Intl. Found. for the Conserv. of Birds, North Hollywood, CA.

Gee, G. F., and T. J. Sexton. 1979. Artificial insemination of cranes with frozen semen. pp. 89–94 *In*: J. C. Lewis (ed.), 1978 Crane Workshop.

Gee, G. F., and S. A. Temple. 1978. Artificial insemination for breeding non-domestic birds. *In*: P. F. Watson (ed.), Symposium on Artificial Breeding of Non-domestic animals. Zool. Soc. (London), 43: 153–173.

Gockel, S. F., and H. G. Lebherz. 1981. "Conformational" isozymes of ascarid enolase. J. Biol. Chem., 256: 3875–3883.

Goodman, M. (ed.). 1982. Macromolecular Sequences in Systematics and Evolutionary Biology. Plenum Press, New York. 418 pp.

Goodman, M., R. E. Tashian, and J. H. Tashian. 1976. Molecular Anthropology: Genes and Proteins in the Evolutionary Ascent of Primates. Plenum Press, New York. 466 pp.

Gorzula, S., C. L. Arocha-Pinango, and C. Salazar. 1976. A method of obtaining blood by caudal vein from large reptiles. Copeia, 1976: 838–839.

Graham, E. F., M. K. L. Schmehl, B. K. Evensen, and D. S. Nelson. 1978. Semen preservation in non-domestic animals. *In*: P. F. Watson (ed.), Symposium on Artificial Breeding of Non-domestic Animals. Zool. Soc. (London), 43: 153–173.

Harris, H., and D. A. Hopkinson. 1976. Handbook of Enzyme Electrophoresis in Human Genetics. North-Holland Publ. Co., Amsterdam.

Hawkes, J. G. (ed.). 1968. Chemotaxonomy and Serotaxonomy. Academic Press, New York. 299 pp.

Hay, R. J. 1978. Preservation of cell-culture stocks in liquid nitrogen. Tissue Culture Assoc. Manual, 4: 787–790.

Hay, R. J. 1979. Identification, separation and culture of mammalian tissue cells. pp. 143–160 *In*: E. Reid (ed.), Cell Populations, Methodology Surveys (B): Biochemistry. Vol. 8. Wiley and Sons, New York.

Hay, R. J., C. D. Williams, M. L. Macy, and K. S. Lavappa. 1982. Cultured cell lines for research on pulmonary physiology available through the American Type Culture Collection. Amer. Rev. Resp. Diseases, 125: 222–232.

Hunziker, J. H. 1969. Molecular data in plant systematics. pp. 280–318 *In:* C. G. Sibley (convenor), Systematic Biology. Publ. 1692, Nat. Acad. Sci. USA, Washington, D.C.

Irwin, H. S., W. W. Payne, D. M. Bates, and P. S. Humphrey (eds.). 1973. America's Systematics Collections: A National Plan. Assoc. of Systematics Collections, Lawrence, Kansas. 63 pp.

Jensen, U., and D. E. Fairbrothers (eds.). 1983. Proteins and Nucleic Acids in Plant Systematics. Springer-Verlag, New York. 408 pp.

Jiminez-Marin, D., and H. C. Dessauer. 1973. Protein phenotype variation in laboratory populations of *Rattus norvegicus*. Comp. Biochem. Physiol., 46B: 487–492.

Jiminez-Porras, J. M. 1961. Biochemical studies on the venom of the rattlesnake *Crotalus atrox atrox*. J. Exper. Zool., 148: 251–258.

Jiminez-Porras, J. M. 1964. Venom proteins of the fer-de-lance, *Bothrops atrox* from Costa Rica. Toxicon, 2: 155–166.

Keilen, D., and Y. L. Wang. 1947. Stability of hemoglobin and certain non-erythrocytic enzymes in vitro. Biochem. J., 41: 491–499.

Kitto, G. B., P. M. Wasserman, and N. O. Kaplan. 1966. Enzymatically active conformers of mitochondrial malate dehydrogenase. Proc. Nat. Acad. Sci. USA, 56: 578–585.

Klebe, R. J. 1975. A simple method for quantitation of isozyme patterns. Biochem. Genet., 13: 805–812.

Kohne, D. E. 1970. Evolution of higher-organism DNA. Quart. Rev. Biophysics, 33: 327–375.

Krzynowek, J., and K. Wiggen. 1981. Genetic identification of cooked and frozen crabmeat by thin layer polyacrylamide gel isoelectric focusing: Collaborative study. J. Assoc. Official Analyt. Chem., 64: 670–673.

Lake, P. E. 1978. The principles and practice of semen collection and preservation in birds. Symp. Zool. Soc. Lond., 43: 31–49.

Lawson, R., and H. C. Dessauer. 1979. Biochemical genetics and systematics of garter snakes of the *Thamnophis elegans-couchii-ordinoides* complex. Occas. Papers Mus. Zool., Louisiana State Univ., 56: 1—24.

Leboffe, M. J. 1979. Protein variation and genetic distance in the genus *Lemur*. Unpubl. M.S. Thesis, San Diego State Univ., San Diego, California. 96 pp.

Lee, W. L., B. M. Bell, and J. F. Sutton. 1982. Guidelines for Acquisition and Management of Biological Specimens. Assoc. of Systematics Collections, Lawrence, Kansas. 42 pp.

Leone, C. A. (ed.). 1964. Taxonomic Biochemistry and Serology. Roland Press, New York. 728 pp.

Leone, C. A. 1968. The immunotaxonomy literature. Serological Museum Bulletin, Rutgers Univ., New Brunswick, New Jersey, 39: 7–24.

Lillevik, H. A., and C. L. Schloemer. 1961. Species differentiation in fish by electrophoretic analysis of skeletal muscle proteins. Science, 134: 2042–2043.

Lowenstein, J. M., V. M. Sarich, and B. J. Richardson. 1981. Albumin systematics of the extinct mammoth and Tasmanian wolf. Nature, 291: 409–411.

Manwell, C., and C. M. A. Baker. 1970. Molecular Biology and the Origin of Species. Univ. Washington Press, Seattle. 394 pp.

Mao, S.H., and H. C. Dessauer. 1971. Selectively neutral mutations, transferrins and the evolution of natricine snakes. Comp. Biochem. Physiol., 40A: 669–680.

Markert, C. L. 1982. Parthenogenesis, homozygosity, and cloning in mammals. J. Hered., 73: 390–397.

Maure, R. R. 1978. Freezing mammalian embryos: A review of techniques. Theriogenology, 9: 45–68.

Mazur, P. 1970. Cryobiology: The freezing of biological systems. Science, 168: 939–949.

Mazur, P. 1980. Fundamental aspects of the freezing of cells with emphasis on mammalian ova and embryos. Ninth Inter. Congr. Anim. Reprod. Artificial Insemination, pp. 99–114.

McWright, C. G., J. J. Kearney and J. L. Mudd. 1975. Effect of environmental factors on starch gel electrophoretic patterns of human erythrocyte acid phosphatase. pp. 151–161 *In* : G. Davies (ed.), Forensic Science ACS Symp. 13, Amer. Chem. Soc., Washington, D.C.

Mellor, J. D. 1978. Fundamentals of Freeze-Drying. Academic Press, New York. 386 pp.

Menzies, R. A. 1981. Biochemical population genetics and the spiny lobster larval recruitment problem: An update. Proc. Gulf Caribbean Fisheries Inst., 33: 230–243.

Miniatis, T., E. F. Fritsch, and J. Sambrook. 1982. Molecular Cloning: A Laboratory Manual. Cold Spring Harbor Laboratory, Cold Spring Harbor, New York.

Moore, D. W., and T. L. Yates. 1983. Rate of protein inactivation in selected animals following death. J. Wildl. Management, 47: 1166–1169.

Munjaal, R. P., T. Chandra, S. L. C. Woo, J. R. Dedman, and A. R. Means. 1981. A cloned calmodulin gene probe is complementary to DNA sequences of diverse species. Proc. Nat. Acad. Sci. USA, 78: 2330–2334.

Myers, N. 1979. The Sinking Ark. Pergamon Press, Elmsford, New York. 307 pp.

Nakanishi, M., A. C. Wilson, R. A. Nolan, G. C. Gorman, and G. S. Bailey. 1969. Phenoxyethanol: Protein preservative for taxonomists. Science, 163: 681–683.

Nei, M. 1975. Molecular Population Genetics and Evolution. North-Holland Press, Amsterdam.

Nevo, E. 1978. Genetic variation in natural populations. Theor. Popul. Biol., 13: 121–177.

Nuttall, G. H. F. 1904. Blood Immunity and Blood Relationships. Cambridge Univ. Press, England.

Parker, W. C., and A. G. Bearn. 1960. Alteration in sialic acid content of human transferrin. Science, 133: 1014–1016.

Polge, C. 1978. Embryo transfer and embryo preservation. *In*: P. F. Watson (ed.), Symposium on Artificial Breeding of Non-domestic Animals. Zool. Soc. (London), 43: 303–316.

Purcel, V. G. 1979. Advances in preservation of swine spermatozoa. *In* : H. Hawk (ed.), Animal Reproduction. Beltsville Agri. Res. Symp., 3: 145–157.

Richardson, B. (chairman). 1982. Report-Workshop on Frozen Tissue Collections in Museums. Bureau of Flora and Fauna, Australia.

Russell, E. S. (chairman). 1978. Conservation of Germplasm Resources: An Imperative. Committee on Germplasm Resources, Nat. Acad. Sci. USA, Washington, D.C. 118 pp.

Russell, F. E., J. A. Emery, and T. E. Long. 1960. Some properties of rattlesnake venom following 26 years of storage. Proc. Soc. Exper. Biol. Med., 103: 737–739.

Ryder, O. A., A. T. Bowling, P. C. Brisbin, P. M. Carroll, I. K. Gadi, S. K. Hansen, and E. A. Wedemeyer. 1983. Genetics of *Equus przewalskii* Polikov 1881: Analysis of genetic variability in breeding lines, comparison of equid DNAs, and a brief description of a cooperative breeding program in North America. Equus (Berlin), Vol. 2. In press.

Ryder, O. A., P. C. Brisbin, A. T. Bowling, and E. A. Wedemeyer. 1981. Monitoring genetic variation in endangered species. pp. 417–424 *In* : G. G. E. Scudder, and J. L. Reveal (eds.), Evolution Today. Hunt Inst. for Botanical Documentation.

Ryder, O. A., R. A. Fisher, W. Putt, and D. Whitehorse. 1982. Genetic differences among subgroups of a captively-bred endangered species: The case of the Mongolian wild horse, *Equus przewalskii*. Proc. Amer. Assoc. Zool. Parks and Aquaria, 1982, pp. 91–102.

Ryder, O. A., and S. K. Hansen. 1979. Molecular cytogenetics of the Equidae. I. Purification and cytological localization of a (G + C)-rich satellite DNA from Przewalskii's horse, *Equus przewalskii*. Chromosoma, 72: 115–129.

Sanger, F. 1981. Determination of nucleotide sequences in DNA. Science, 214: 1205–1210.

Sanger, F., S. Nicklen, and A. R. Coulson. 1977. DNA sequencing with chain-terminating inhibitors. Proc. Nat. Acad. Sci. USA, 74: 5463–5467.

Sarich, V. M. 1977. Rates, sample sizes, and the neutrality hypothesis for electrophoresis in evolutionary studies. Nature, 265: 24–28.

Seager, S., D. Wildt, and C. Platz. 1978. Artificial breeding of non-primates. *In*: P. F. Watson (ed.), Symposium on Artificial Breeding of Non-domestic Animals. Zool. Soc. (London), 43: 207–218.

Seidel, G. E. J. 1981. Superovulation and embryo transfer in cattle. Science, 211: 351–358.

Selander, R. K., and W. E. Johnson. 1973. Genetic variation among vertebrate species. Ann. Rev. Ecol. Syst., 4: 75–91.

Senner, J. W. 1980. Inbreeding depression and the survival of zoo populations. pp. 209–229 *In*: M. F. Soule, and B. A. Wilcox (eds.) Conservation Biology. Sinauer Publ., Sunderland, Maryland.

Sensabaugh, G. F., A. C. Wilson, and P. L. Kirk. 1971. Protein stability in preserved biological remains. I. Survival of biologically active proteins in an 8-year-old sample of dried blood. International J. Biochem., 2: 545–557.

Sensabaugh, G. F., A. C. Wilson, and P. L. Kirk. 1971. Protein stability in preserved biological remains. II. Modification and aggregation of proteins in an 8-year-old sample of dried blood. International J. Biochem., 2: 558–568.

Sexton, T. J. 1976. Studies on the fertility of frozen fowl semen. VIII Int. Congr. Anim. Reprod. Artificial Insemination (Krakow), 4: 1079–1982.

Sexton, T. J., and G. F. Gee. 1978. A comparative study on the cryogenic preservation of semen from the sandhill crane and the domestic fowl. *In*: P. F. Watson (ed.), Symposium on Artificial Breeding of Non-domestic Animals. Zool. Soc. (London), 43: 89–95.

Shah, D. M., and C. H. Langley. 1979. Inter- and intraspecific variation in restriction maps of *Drosophila* mitochondrial DNAs. Nature, 281: 696–699.

Shields, G. F., and N. S. Straus. 1975. DNA-DNA hybridization studies of birds. Evolution, 29: 159–166.

Shields, W. M. 1979. Philopatry, Inbreeding, and the Adaptive Advantages of Sex. Unpubl. Ph.D. diss., Ohio State Univ., Columbus, Ohio.

Sibley, C. G. (Convenor). 1969a. Systematic Biology: Proc. International Conf., Nat. Acad. Sci. USA, Publ. 1692. 632 pp.

Sibley, C. G. 1969b. A report on Program B of the ALPHA HELIX expedition to New Guinea. Discovery, 5: 39–46.

Sibley, C. G., and J. E. Ahlquist. 1981a. Instructions for specimen preservation for DNA extraction: A valuable source of data for systematics. Assoc. of Systematics Collections Newsletter, 9: 44–45.

Sibley, C. G., and J. E. Ahlquist. 1981b. The phylogeny and relationships of ratite birds as indicated by DNA-DNA hybridization. pp. 301–335 *In*: G. G. E. Scudder, and J. L. Reveal (eds.), Evolution Today, Proc. Second Inter. Congr. Syst. Evol. Biol. Hunt Inst. Botanical Documentation.

Sibley, C. G., and J. E. Ahlquist. 1983. Phylogeny and classification of birds based on the data of DNA-DNA hybridization. pp. 245–292 *In*: R. F. Johnston (ed.), Current Ornithology, Vol. 1. Plenum Press, New York.

Sigman, D. S., and M. A. B. Brazier. 1980. The Evolution of Protein Structure and Function. UCLA Forum in Med. Sci., 21: 1–350. Academic Press, New York.

Smith, M. H., M. W. Smith, S. L. Scott, E. H. Liu, and J. C. Jones. 1983. Rapid evolution in a post-thermal environment. Copeia, 1983: 193–197.

Smith, M. W., C. F. Aquadro, M. H. Smith, R. K. Chesser, and W. J. Etges. 1982. Bibliography of Electrophoretic Studies of Biochemical Variation in Natural Populations. Texas Tech Press, Lubbock, Texas.

Smith, M. W., M. H. Smith, and R. K. Chesser. 1983. Biochemical genetics of mosquitofish. I. Environmental correlates and temporal and spatial heterogeneity of allele frequencies within a river drainage. Copeia, 1983: 182–193.

Southern, E. M. 1975. Detection of specific sequences among DNA fragments separated by gel electrophoresis. J. Molec. Biol., 98: 503–517.

Stuessy, T. F., and K. S. Thomson (eds.). 1981. Trends, Priorities and Needs in Systematic Biology. Assoc. of Systematics Collections, Lawrence, Kansas. 51 pp.

Svedberg, T. 1934. The sedimentation constants of the respiratory proteins. Biol. Bull., 66: 191–223.

Taylor, H. A., S. E. Riley, S. E. Parks and R. E. Stevenson. 1978. Longterm storage of tissue samples for cell culture. In Vitro, 14: 476–478.

Terborgh, J., and B. Winter. 1980. Some causes of extinction. pp. 119–133 *In*: M. F. Soule, and B. A. Wilcox (eds.), Conservation Biology. Sinauer Publ., Sunderland, Maryland.

Wake, D. B. (chairman). 1975. Report of the committee on resources in herpetology. Copeia, 1975: 391–404.

Waladsen, S. M., A. O. Trounson, and L. E. A. Rowson. 1977. Transplantation of sheep and cattle embryos after storage at minus 196°celsius. pp. 190–194 *In*: K. Elliott, and J. Whelan (eds.), Ciba Foundation Symposium Vol. 52, London.

Wickham, A. (conference coordinator). 1981. Proceedings of the U.S. Strategy Conference on Biological Diversity. U.S. Dept. of State Publ. 9262, Washington, D.C. 126 pp.

Wilkens, N. P., D. O'Regan, and E. Moynihan. 1978. Electrophoretic variability and temperature sensitivity of phosphoglucoisomerase and phosphoglucomutase in Littorinids and other marine mollusks. pp. 141-155 *In*: B. Battaglia, and J. A. Beardmore (eds.), Marine Organisms Genetics, Ecology and Evolution. Plenum Press, New York.

Willett, W. C., and B. MacMahon. 1984. Diet and cancer: An overview. New England J. Med., 310: 633–638.

Wilson, A. C., S. S. Carlson, and T. J. White. 1977. Biochemical Evolution. Ann. Rev. Biochem., 46: 573–639.

Wright, C. A. (ed.). 1974. Biochemical and Immunological Taxonomy of Animals. Academic Press, New York. 490 pp.

Yardley, D., J. C. Avise, J. W. Gibbons, and M. H. Smith. 1974. Biochemical genetics of sunfish. III. Genetic subdivision of fish populations inhabiting heated waters. pp. 255–263 *In*: J. W. Gibbons, and R. R. Sharitz (eds.), Thermal Ecology. AEC Symp. Series CONF-73055.

Yoshida, A. 1966. Glucose-6-phosphate dehydrogenase of human erythrocytes. I. Purification and characterization of normal (B+) enzyme. J. Biochem., 241:4966–4976.

SECTION VII

AMERICA'S COLLECTIONS OF FROZEN TISSUES: RECOMMENDATIONS FOR A NATIONAL PLAN

The conservation of biological diversity and the maintenance of unpolluted environments are critical problems of our time (Myers, 1979). Species of plants and animals are disappearing at accelerating rates, and an ever-increasing variety of industrial pollutants is making it more and more difficult to maintain air, land, and water environments supportive of living forms. The Committee on Germplasm Resources of the Natural Research Council (Russell, 1978), participants in the Strategy Conference on Biological Diversity sponsored by the U.S. Department of State (Wickham, 1982), and international agencies in the U.S.S.R. (Academy of Sciences, U.S.S.R., 1980) and Australia (Richardson, 1982) have recognized these problems and offered general recommendations designed to solve them. The cryopreservation of tissue samples from both wild and domesticated organisms figures as a prime recommendation in the reports stemming from each of these conferences.

Collections of frozen tissues taken from wild and domesticated organisms are at present the most practical and economical means of fulfilling this recommendation. Such material offers a retrievable source of the unique genetic material of individual organisms and is thus useful for addressing many problems concerned with the preservation of biological diversity. In addition, these collections provide materials for basic research studies in systematics, genetics, developmental biology, virology, biochemistry, and immunology focused at the molecular level. Tissues in these collections (and antisera raised to proteins of the tissues) are of economic importance to scientists involved in breeding programs with endangered species, veterinarians caring for animals in zoological parks, technicians monitoring levels of environmental pollutants, and police, wildlife personnel, and customs agents confronted with forensic problems involving endangered species and regulated game. There is little doubt that in the future, additional uses in both basic and applied research will be made of frozen tissue collections.

Many scientists throughout the world have already acquired large holdings of frozen tissues from a wide variety of living organisms (Dessauer and Hafner, 1984). Nearly all of these tissue banks began as a direct extension of the research interests of the scientist-in-charge; as the existence of these collections became more widely known, their holdings were put to an ever-increasing variety of uses. Nevertheless, funding to maintain these collections has remained precarious, dependent in most cases on basic research grants to the individual scientist. Irreplaceable materials have been lost for the want of funds to maintain a proper curatorial program and purchase, and maintain, ultracold freezers. Valuable materials have been discarded upon termination of research projects because of the impracticability of storing them for and disseminating them to other potential users. Today, the large amounts of tissues stored in freezers across the nation constitute an unmanaged, largely unrecognized, national resource.

Prompted by the need for a national plan to coordinate, manage, and fund frozen tissue collections, a panel was convened in May of 1983 to discuss the present state of American frozen tissue collections and develop such a plan. This meeting was titled "Workshop on Frozen Tissue Collection Management" and was funded by the National Science Foundation and sponsored by the Association of Systematics Collections. Following are the recommendations of this panel.

RECOMMENDATIONS

1. ESTABLISH A COUNCIL OF THE ASSOCIATION OF SYSTEMATICS COLLECTIONS (ASC) CHARGED WITH IMPLEMENTATION OF THE NATIONAL PLAN FOR FROZEN TISSUE COLLECTIONS.

   This Council should be broadly representative of the community of frozen tissue collection managers, and charged as per Recommendation 3 and Priority A-4 of *America's Systematics Collections: A National Plan* (Irwin, 1973). This council "will collect data concerning resources in the disciplines, develop criteria for distinguishing National Resources, recommend recognition of appropriate collections as National Resource Centers, and stimulate discipline-wide studies of ways to improve the condition and services of systematics collections" (Irwin, 1973, p. 29). This Council should work closely with program officers of the National Science Foundation and National Institutes of Health concerned with the development and preservation of national biological resources.

2. A LIMITED NUMBER OF EXISTING FROZEN TISSUE COLLECTIONS IN THE UNITED STATES SHOULD BE DESIGNATED AS *FROZEN TISSUE DEPOSITORIES,* CHARGED WITH ACCEPTING, PRESERVING, AND DISTRIBUTING TISSUES FOR THE SCIENTIFIC COMMUNITY AT LARGE.

   This designation should be conferred only on those collections that meet specific requirements. A proposal should be submitted to an appropriate federal granting agency (see Recommendation 4) that describes the facility, its existing staff and collection holdings, and provides tangible evidence of the institution's commitment to long term financial support for the depository. The curator must provide convincing evidence that, if the collection at his institution is selected as a *Frozen Tissue Depository* supported by federal funds, he will accept, preserve, and distribute tissues for the scientific community at large. This proposal should document a clear commitment to service, with funds requested solely for collection curation.

   Of the larger existing collections, those located at the following universities would seem to be prime candidates (in alphabetical order by state): University of California at Berkeley, University of Illinois at Urbana, Louisiana State University, Wayne State University, University of Minnesota, University of New Mexico, and Texas Tech University. In addition, the collection maintained by the Zoological Society of San Diego would seem to be an appropriate candidate because of its size, quality, and the nature of its research commitment.

3. COLLECTIONS DESIGNATED AS *FROZEN TISSUE DEPOSITORIES* SHOULD, IN GENERAL, BE MANAGED IN ACCORDANCE WITH TRADITIONAL CURATORIAL PROCEDURES FOR SYSTEMATIC COLLECTIONS.

The time-tested curatorial practices currently employed in American systematic collections should be scrupulously followed by curators of frozen tissue collections. A collection catalog must be maintained. Each specimen entered into the collection catalog should be documented as to species, sex, exact locality of origin, date of capture, nature of tissues taken, location of voucher specimen, and other pertinent information. Whenever possible, a specimen incorporated into the collection should be accompanied by a traditionally preserved voucher specimen (skin, skeleton, fluid specimen) deposited in an accredited systematic collection.

Often, the principal contribution to a molecular study is the field work of the naturalist who provides the tissue upon which the study is based. Therefore, the person donating the tissues to the depository should be suitably recognized in any publication resulting from their use. Costs for shipment of tissues from the depository should be paid by the recipient, as such material will normally be destroyed during analysis and is, therefore, a *gift* rather than a loan.

Collections of frozen tissues should be managed using a computerized inventory system. Such a system will provide many benefits including ready access to tissue samples (reducing the time spent searching for samples in freezers) and increased flexibility in record keeping necessary to handle constant deacquisition of materials.

4. MAJOR FINANCIAL SUPPORT FOR DESIGNATED *FROZEN TISSUE DEPOSITORIES* SHOULD BE SHARED BY THE HOME INSTITUTION AND THE BIOLOGICAL RESOURCE PROGRAMS OF THE NATIONAL SCIENCE FOUNDATION AND THE NATIONAL INSTITUTES OF HEALTH.

The home institution should provide the physical facilities (space, power, and water) and service functions (secretarial and janitorial) for the *Frozen Tissue Depository*. The home institution should also be expected to demonstrate a long term commitment to its collection by funding the salary of a full time professional Curator of Frozen Tissues.

The biological resource support programs of both the National Science Foundation and the National Institutes of Health should share costs for the purchase of capital equipment (e.g., ultracold freezers, alarm systems), curatorial supplies (e.g., cryotubes, liquid nitrogen), equipment repair, and shipment of tissues to the depository. These agencies should provide the salary for at least one support person in each *Frozen Tissue Depository*.

Because of the broad value of the frozen tissue resource, many funding agencies, in addition to NSF and NIH, should participate in the development of *Frozen Tissue Depositories*. This participation will be mostly indirect, through support of research concerned with cryopreservation techniques and funding of field projects that contribute tissues to collections. Funding agencies should insist that investigators utilize the network of depositories whenever possible.

5. REGULATORY AGENCIES OF THE U.S. GOVERNMENT SHOULD PROMOTE THE INTERNATIONAL EXCHANGE OF FROZEN TISSUES.

Agencies regulating importation, exportation and domestic transport of biological materials must recognize the value of frozen tissues for basic and applied research and take steps to simplify regulations for tissue importation, exportation and transport.

The Workshop Panel members, and most scientists, agree that federal regulations concerning the importation of biological materials are needed to protect domestic crops and livestock, safeguard the general public health, and control illicit trafficking in endangered and threatened species. However, these regulations, though necessary, should not be self-defeating by inhibiting *bona fide* medical, veterinary, and basic research. Scientists who import frozen tissues now constitute an increasingly large portion of the research community, and their use of these tissues should not automatically be considered a threat to the health and welfare of the country. The Workshop Panel recommends, therefore, that the regulatory agency concerned, the United States Department of Agriculture, implement swift, constructive measures to reevaluate its policies in regard to the importation of frozen tissues.

6. INVENTORIES OF COLLECTION HOLDINGS AND ADVANCES IN THE FIELDS OF TISSUE CRYOPRESERVATION AND TISSUE COLLECTION CURATION SHOULD BE PUBLICIZED FOR THE BENEFIT OF ALL.

The ASC Newsletter would be an appropriate outlet for this kind of information. Curators of designated *Frozen Tissue Depositories* should be encouraged to publish a list of collection holdings with periodic updates.

LITERATURE CITED

Academy of Sciences, USSR. 1980. Genome Conservation. Moscow, USSR.

Dessauer, H. C. and M. S. Hafner (eds.). 1984. Collections of Frozen Tissues: Value, Management, Field and Laboratory Procedures, and Directory of Existing Collections. Association of Systematics Collections, Lawrence, KS.

Irwin, H. S., W. W. Payne, D. M. Bates and P. S. Humphrey. 1973. America's Systematics Collections: A National Plan. Association of Systematics Collections, Lawrence, KS, 63 pp.

Myers, N. 1979. The Sinking Ark. Pergamon Press, Elmsford, NY, 307 pp.

Richardson, B. (chairman). 1982. Workshop on Frozen Tissue Collections in Museums (report). Australian Bureau of Flora and Fauna.

Russell, E. S. (chairman). 1978. Conservation of Germplasm Resources: An Imperative. National Academy of Sciences, Washington D.C., 118 pp.

Wickham, A. (conference coordinator). 1981. Proceedings of the U.S. Strategy Conference on Biological Diversity. U.S. Department of State, Publ. 9262, Washington, D.C., 126 pp.